AF566753

COSMO COMPANIONS

COSMO

DICTIONARY OF ECOLOGY

COSMO COMPANIONS

COSMO DICTIONARIES comprise of quality reference works which are comprehensive, latest, and published with finest production techniques, compiled by a trained team of experts especially for this series.

COSMO DICTIONARY OF ABBREVIATIONS

COSMO DICTIONARY OF FINANCE

COSMO DICTIONARY OF BUSINESS

COSMO DICTIONARY OF ANTHROPOLOGY

COSMO DICTIONARY OF LITERARY TERMS

COSMO DICTIONARY OF PHILOSOPHY

COSMO DICTIONARY OF EDUCATION

COSMO DICTIONARY OF LITERATURE

COSMO DICTIONARY OF PLANT NAMES

COSMO DICTIONARY OF PSYCHOLOGY

COSMO DICTIONARY OF ASTROLOGY

COSMO DICTIONARY OF POLITICS

COSMO DICTIONARY OF SOCIOLOGY

COSMO DICTIONARY OF MATHEMATICS

COSMO DICTIONARY OF BIOLOGY

COSMO DICTIONARY OF PHYSICS

COSMO DICTIONARY OF CHEMISTRY

COSMO DICTIONARY OF GEOGRAPHY

COSMO DICTIONARY OF EDUCATION

COSMO DICTIONARY OF EUROPEAN LITERATURE

COSMO

DICTIONARY OF ECOLOGY

By

HERBERT C. HANSON

The Catholic University of America
Washington, D.C.

COSMO PUBLICATIONS

2008 **New Delhi**

PUBLISHED BY
COSMO PUBLICATIONS,
24-B, Ansari Road, Darya Ganj,
New Delhi 110 002, India.

COSMO DICTIONARY OF ECOLOGY

First COSMO Edition 2008

ISBN 81-307-0659-8

PRINTED AND BOUND IN INDIA

ABBREVIATIONS

cf.—confer, compare this definition with the definition of words that follow.

q. v.—quod vide, indicating that it is desirable to look up the definition of the preceding word in order to understand more fully the definition being considered.

Syn.—Synonym.

Italicized words indicate that they are defined in this book, or that such words form the scientific name of a plant or animal.

PREFACE

The aim of this dictionary is to fill the need for definitions of many new terms that have come into usage during the past thirty years and also to include many of the old terms that are used in current literature. It has not seemed desirable to include many words that are rarely if at all used at present and which are in the older glossaries. Many words from fields closely related to ecology such as forestry, range management, agronomy, soils, and genetics are included because of their wide usage in ecological literature.

The definitions are usually those that are in accord with present general usage. It has not seemed wise to attempt to pronounce judgment on the desirability of, or the need for, certain terms since such decisions are made by usage as a language grows. The inclusion of certain words in this dictionary does not necessarily imply that the author approves or recommends their use.

Words are tools of thought. Clarification of the meaning of terms, precision in their use, and uniformity in usage among workers in ecology and related fields are essential in the growth of a science. When a concept or process can be expressed precisely in ordinary language it appears unnecessary, and indeed detrimental to the growth of a science, to coin a new term. It is worth while to make ecological literature intelligible to as wide a field of readers as possible. As concepts and techniques become clearer and more precise new terms are often needed, and when a new word is accepted in one branch of science it should be accepted in other branches.

Definitions are not immutable. As knowledge increases the meanings of terms change. It is hoped that this volume will aid in the development of an increasingly useful ecological terminology.

A list of references are given below, many of which will be found useful in securing further amplification of the meaning of words in this dictionary.

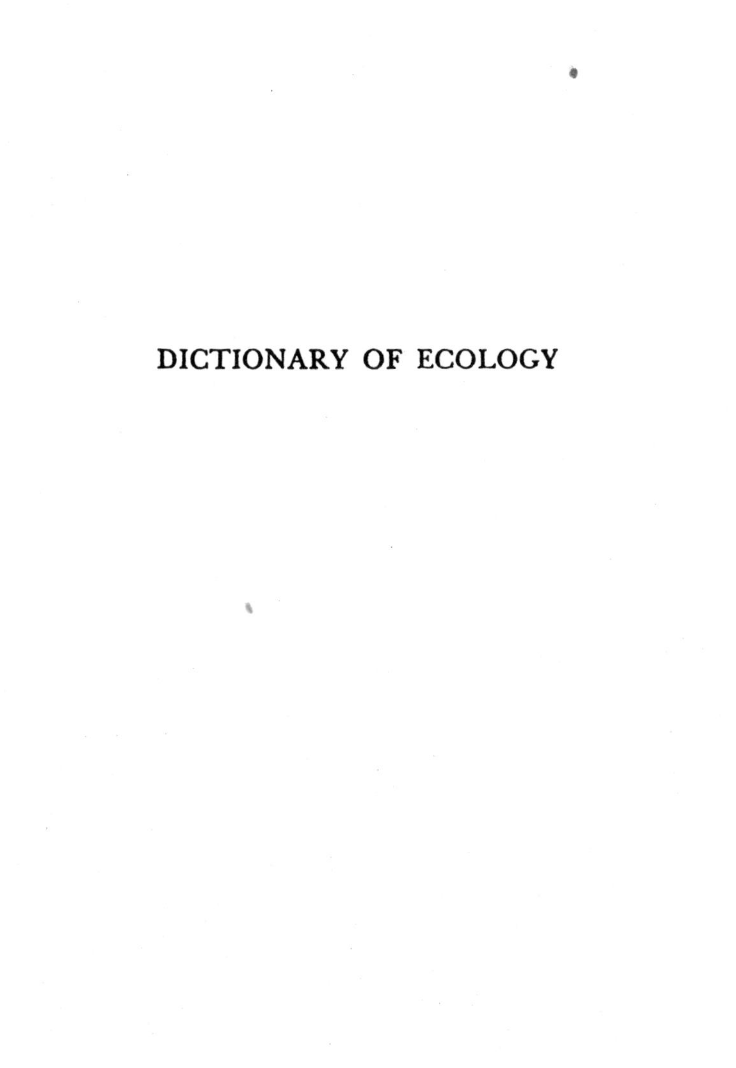

DICTIONARY OF ECOLOGY

A

Aapamoor

A mosaic of high moor and low moor, consisting of circular or elongated mounds covered with dwarf shrubs and sphagnum and depressions occupied by mostly sedges and sphagnum.

Abicoen

The non-biotic elements of a habitat.

Abioseston

See *Seston*.

Abrasion Platform

The part of the continental shelf and terrace on which a horizontal plain is formed by long continued wave action.

Absolute Extremes

The highest (absolute maximum) and lowest (absolute minimum) values of a meteorological element, especially temperature, that have ever been recorded at a station.

Absolute Humidity

See *Humidity, absolute*.

Absorption Loss

The initial loss of water from a canal or reservoir by capillary action and percolation.

Abstract Community

A generalized category comprising a number of similar units or stands of vegetation and including animal life.

Abundance

The total number of individuals of a species in an area, population, or community. The *index of relative abundance* gives a useful approximation of numbers relative to time or space, e.g., the number of rabbits seen in an hour in a certain place. cf. *Density*.

Abyssal

Refers to the deepest region of the ocean and often used for the zone in lakes below the *profundal* (q. v.). See *Archibenthal*.

Abyssal-benthic (Abyssobenthic)

The lower part of the *abyssal* region, below about 3300 feet (1000 meters).

Acarophytium

The symbiotic relationship of mites and plants.

Acaulescent

Refers to a plant with inconspicuous, or seemingly absent, stem above the ground.

Accelerated Erosion

Washing away or blowing away of soil material in excess of *normal erosion* (q. v.), resulting from changes in the vegetation cover or ground conditions.

Accidental

A species that occurs with a low degree of *Fidelity* in a *Community-type*.

Acclimation

The increased tolerance or physiological adjustment of an organism to a change in its environment.

Accommodation

Modification of the focus of the eye.

Accrescent

Refers to plants which continue to grow after flowering.

Accumulator Plant

A plant that absorbs certain elements which accumulate in its tissues to a much higher degree than in most plants, e.g., *Equisetum* concentrates large amounts of silica. cf. *Seleniferous.*

Acheb

A short-lived type of vegetation, characterized chiefly by mustards and grasses, in the Italian Sahara.

Achene

A one-seeded, dry, non-opening fruit in which the seed is not attached to the wall of the fruit, e.g., sunflower "seed."

Aciculignosa

Coniferous vegetation with small, evergreen, needle-like leaves.

Acid Bog

See *Bog.*

Acidophilous

Refers to organisms that grow well or exclusively on soil or in a medium that is acid in reaction.

Acid Soil

A soil with an acid reaction, containing more hydrogen than hydroxyl ions; for practical purposes with a pH below 6.6.

Acidotrophic

Refers to organisms that feed on food having an acid reaction.

Acquired Character

A modification of structure or function appearing during the lifetime of an individual, caused by environmental conditions, e.g., broad leaves of some plants growing in shade.

Acre-Foot

The quantity of water that will cover one acre one foot deep.

Acre-Inch

The quantity of water that will cover one acre one inch deep.

Acrodomatia

Structures on plants that shelter mites.

Acropetal

Refers to the development of organs in plants, the oldest at the base, the youngest at the tip. cf. *Basipetal.*

Acrophytia

Plant communities in alpine regions.

Actic

Refers to the part of the seashore between tide marks. See *Littoral.*

Actinometer

An instrument which measures radiant energy, especially the property that produces chemical effects.

Actinometry

The measurement of chemical reactions caused by radiation.

Actinomycetes

A group of organisms possessing very fine hyphae or

threads, classified with bacteria or fungi. Various kinds cause decomposition, disease, or produce antibiotics such as *streptomycin* (q.v.).

Action

The impingement of environmental factors such as heat or light upon organisms.

Actium

A plant-animal community on a rocky seashore.

Activated Sludge

Material composed chiefly of bacteria and protozoa, used in one method of sewage disposal.

Actophilous

Refers to organisms that grow well on rocky seashores.

Adaptability

Capability of an organism to make changes which fit it better to its environmental conditions.

Adaptable

Refers to the adaptability of an organism.

Adaptation

(1) The process or processes by which an organism becomes apparently better suited to its environment or for particular functions. (2) The structures or activities of an organism, or of one or more of its parts, which tend to fit it better for life in its environment. (3) The adapted form.

Adaptedness

The sum of genetic characters by which an organism is suited to its environment.

Adaption

See *Adaptation*.

Adaptive Radiation

The evolution of *taxa* (q. v.) as they become adapted to

new habitats, applicable also to the development of a new community.

Adaptive Selection
The evolution of more or less similar forms in separate but ecologically similar areas.

Additive
A material added to a fertilizer, or to another substance, to improve its chemical or physical condition.

Adiabatic
Refers to an occurrence in which heat is neither gained nor dissipated.

Adjustment
Processes by which an organism becomes better fitted to its environment; functional, never structural. cf. *Adaptation.*

Adobe
A fine calcareous clay or silt, may be mixed with water to make bricks for construction purposes. cf. *terron.*

Adsorption
The attachment of molecules or ions to surfaces or interfaces such as solid-liquid, solid-gas, and liquid-gas boundaries.

Advance Growth
Young trees in openings or under the canopy in forests before cutting or regeneration operations are started. syn. Advance reproduction. cf. *Second growth.*

Adventicous Species
Organisms which have invaded from a distance.

Adventitious
An organ growing out of its usual location, e.g., root from a stem; a species which has invaded from another area and has become more or less naturalized.

Adventive
A plant growing spontaneously, not native, ephemeral or not spreading appreciably.

Aelophilous
Refers to dissemination by wind.

Aeolian
Refers to the wind, or to soil materials which have been moved by the wind or are subject to such movement.

Aeration
The processes by which air and other gases in a medium are renewed or exchanged.

Aerobe
An organism capable of living only in the presence of free oxygen.

Aerenchyma
Tissue with thin-walled cells and large air spaces, especially common in aquatic plants.

Aerial Photograph
A vertical or oblique photograph taken from the air.

Aerobic
Refers to life or a process occurring only in the presence of free oxygen. See *Aerobe*.

Aerobiosis
Life in the presence of free oxygen in the medium.

Aerophyte
See *Epiphyte*.

Aeroplankton
Microorganisms floating in the air, cf. *Plankton*.

Aerotaxis
Involuntary response of an organism to a gas, such as

the curving of a plant toward a higher concentration of oxygen.

Aestidurilignosa

Woodland characterized by mixed evergreen-deciduous hardwoods.

Aestilignosa

Forest or woodland in which the woody plants are leafless in winter and buds are protected by scales, e.g., beech forest.

Aestival

See *Estival.*

Affinity

The relationship between organisms that indicates a common origin; used occasionally to denote similarity of communities.

Afforestation

The process of establishing a forest on an area, especially where forest was not present previously, cf. *reforestation.*

Aftermath

The regrowth of plants after mowing.

After-ripening

The dormancy period, following formation of the seed that is required for changes in the embryo to occur prior to germination.

Agameon

See *Agamospecies.*

Agamic

Asexual.

Agamogenesis

Asexual reproduction, parthenogenesis (q. v.).

Agamospecies

An aggregation of individuals in which reproduction occurs almost exclusively by asexual means. syn. Agameon, binom.

Agamospermy
Production of seed asexually, exclusive of vegetative reproduction, cf. apomixis.

Age and Area
Willis' hypothesis that the older a species is, the larger is the area that it occupies.

Age Class
A stand in which all of the trees originated in the same regeneration period. cf. *Even-aged.*

Age Distribution
The classification of individuals of a population according to age classes or periods such as prereproductive, reproductive, and postreproductive.

Agglutination
Formation of clumps of microorganisms or cell inclusions.

Aggradation
Building up of a portion of the earth's surface toward uniformity of grade by deposition, as on the bottom of a lake.

Aggregate
A cluster of particles as a crumb of soil; to collect particles into a cluster.

Aggregation
The coming together of organisms into a group, e.g., offspring clustered about the parents. The condition of being collected into a cluster or group. cf. *Community.*

Agonistic Behaviour
An activity such as fighting, feigning, and escaping, connected with conflict between animals.

Agrarian Zone
The portion of a country that can be cultivated.

Agrestal
Refers to an organism that grows in arable ground.

Agrology
The study of soils. See *Edaphology*.

Agronomy
The study of the production, processing, and use of farm crops.

Agrophilous
Refers to organisms that grow well in grain fields or other areas resulting from man's activities.

Agrostology
The branch of systematic botany dealing with grasses.

A Horizon (Soil)
The stratum of soil consisting of one or more of the following layers. (A_0 horizon, partly decomposed or matted plant remains lying on top of the mineral soil. A_{00} horizon, the relatively fresh leaves and other plant debris, generally of the past year, lying on the A_0 horizon.) A_1 horizon, the surface mineral layer, relatively high in organic matter, usually dark in color. A_2 horizon, below the A_1 horizon, in places the surface layer, usually lighter in color than the underlying horizon, in which leaching of solutes and suspended materials occurs. A_3 horizon, transitional to the B horizon, more like A than B, sometimes absent.

Aiphyllus
Evergreen.

Air layering
A method for producing roots on a stem in an aerial position.

Air-sacs

Thin-walled structures, containing air, in birds and in some insects.

Alar, Alary, Alate

Winged.

Albinism

Complete or almost complete absence of pigment, resulting in plants that are white in whole or in part, and in animals with milky-white skin, light hair, and red pupils in eyes.

Albino

An organism deficient in pigment.

Allee's Principle

The extent of aggregation and the degree of density of a population most favorable for optimum growth and survival varies according to the species and environmental conditions. Either deficiency or excess may be detrimental.

Aletophyte

A weed growing in a *mesic* (q. v.) habitat.

Algae

The simplest kind of green plants, usually growing in water or damp places, consisting of several phyla, formerly classified in the *Thallophytes* (q. v.).

Algoid

Resembling an alga.

Algology

The study of *Algae*.

Alien

An introduced plant which has become naturalized.

Aliquote

The constant of temperature for a certain stage in the life-cycle of an organism. See *Temperature summation.*

Alkali Reserve

The total amount of dissolved salts or other substances which tend to maintain the normal alkalinity of a natural water or the internal body fluid of an organism.

Alkali Soil

A soil that has such a high degree of *Alkalinity* (pH 8.5 or higher), or such a high percentage of exchangeable sodium (15 per cent or more), or both, that the growth of most crop plants is reduced or prevented. See *Black alkali, Saline soil.*

Alkaline Soil

A non-acid soil which contains more hydroxyl ions than hydrogen ions; precisely, a soil with pH 7.0 or higher; for practical purposes, with pH 7.3 or higher.

Alkalinity

The chemical state of water or other substance in which the hydroxyl ions exceed the hydrogen ions, usually with pH 7.0 or higher. cf. *Salinity.*

Alleghanian Life Zone

One of the divisions of Merriam's *Austral life zone* (q. v.), east of the 100th meridian. See *Life zone.*

Allelarkean Society

An independent, dense, fixed, civilized society. cf. *Autarkean society.*

Allele

One of the two forms of a gene located at a certain position (locus) on a *homologous chromosome* (q. v.). If one allele of a pair is dominant to the other it largely controls the character, e.g., greenness is dominant over *albinism* (q. v.) in seedlings.

Allelomimetic Behaviour
Two or more animals, mutually stimulated, acting similarly. See *Mimetic.*

Allelomorph
See *Allele.*

Allelopathy
Influence of plants, exclusive of microorganisms, upon each other, caused by products of metabolism.

Allen's Principle (Rule)
Appendages of animals tend to be shorter in cold regions, resulting in reduced loss of heat. Cf. *Bergmann's principle.*

Allergen
A substance which induces *allergy,* or causes symptoms to show. e.g., *Pollen.*

Allergy
Sensitivity resulting in pathologic condition in certain people to substances such as pollen, food, hairs; or may be caused by mental or environmental conditions.

Alliance
A group of plant associations according to Braun-Blanquet classified together on the basis of similarity in floristic and sociological characteristics. See *Association.*

Allochoric
Refers to a species occurring in two or more similar communities in the same region.

Allochthonous
Refers to deposits of material that originated elsewhere, e.g., drifted plant materials on the bottom of a lake. cf. *Autochthonous.*

Allogamy
Cross-fertilization (q. v.). See *Outbreeding.*

Allogenic Succession

The kind of *succession* (q. v.) in which one kind of community replaces another because of a change in the environment which was not produced by the plants themselves, e.g., decrease in soil moisture by improved drainage. cf. *Autogenic succession.*

Allometry

Relationships in the evolutionary development of organs or other characters of organisms which may bring about disharmony, e.g., disproportionate development of antlers and neck muscles of deer or moose.

Allopatric

Refers to organisms originating in or occupying different geographic areas. cf. *Sympatric.*

Allopolyploid

A *polyploid* (q. v.) which originated by the addition of unlike sets of chromosomes. cf. *Autopolyploid, Amphiploid.*

Allopelagic

Refers to organisms occurring at any depth in the sea.

Allotrophic

Refers to lakes or ponds that receive organic material by drainage from the adjacent land. cf. *Autotrophic.*

Alluvial Fan

A fan-shaped deposit of sand, gravel, and fine material from a stream where its gradient lessens abruptly.

Alluvial Soil

Soil that has developed from transported and relatively recently deposited material (*alluvium*), characterized by little or no modification of the original material by soil-forming processes.

Alluvium

Sediments, usually fine materials, deposited on land by a stream.

Alm

A high mountain meadow, alpine or subalpine.

Alm's Fb Coefficient

The relationship of fish caught per hectare to the live weight of the bottom fauna per hectare.

Alpage

A mountain or upland pasture of natural plants grazed by animals at the height of summer. See *Alm*.

Alpha Particle, α Particle

A helium nucleus, given off by nuclei of certain radioactive substances.

Alpha Radiation

A kind of *ionizing radiation* (q. v.) in which alpha particles are given off.

Alpha Ray

A stream of *alpha particles* (q. v.) cf. *Beta ray*.

Alpine

Refers to parts of mountains above tree growth or to organisms living there.

Alternating Communities

See *Twin communities*.

Alternation of Generations

The alternation of different forms in the life cycle of an organism, especially a sexual (gamete-producing) form with a non-sexual (spore-producing) form; occurs in most plants and in many animals.

Alternes

Two or more communities alternating with each other in a more or less restricted area.

Altherbosa

Communities with tall herbs, especially in denuded forest areas.

Altimeter

An instrument for determining altitude, e.g., aneroid barometer.

Altricial

Refers to the condition of delay after birth or hatching in the attainment of a completely independent mode of self maintenance. cf. *Precocial.*

Alvar

A vegetation type consisting of dwarf shrubs resembling steppe, in Sweden.

Alveolate

Pitted, appearing like a honeycomb.

Amanthophilous

Refers to organisms living in sandy areas.

Ambulatorial

Refers to adaptations for walking in contrast to running, associated with forest animals. cf. *Fossorial, Scansorial.*

Amendment

Any material such as lime or synthetic conditioners that is worked into the soil to make it more productive; usually restricted to materials other than fertilizers.

Amensalism

The state or interaction in which one organism is inhibited while the other is not influenced. cf. *Commensalism.*

Ament
A pendulous, spike-like cluster of flowers as in the oaks, willows, and birches. syn. *Catkin.*

Amentaceous
Refers to plants with aments or catkins.

Ametoecious
Refers to a parasite that is restricted to a single host.

Amino Acid
A class of organic compounds containing nitrogen, large numbers of which become linked together to form proteins (q. v.); each one containing at least one amino group ($-NH_2$) and at least one carboxyl ($-COOH$) group.

Amitosis
Direct division of the nucleus of the cell without mitosis (q. v.).

Ammate
A compound of ammonium sulfate used as an herbicide (q. v.).

Ammocolous
Refers to an organism that grows in sand.

Ammonification
The formation of ammonium compounds from organic materials containing nitrogen.

Ammophilous
See *Ammocolous.*

Amnicolous
Refers to organisms inhabiting sandy banks of rivers.

Amniote
An animal such as reptile, bird, or mammal the embryos of which develop within a fluid-filled sac.

Amoeba
A *Protozoan* (q. v.) in the genus *Amoeba.*

Amoebocyte
A cell possessing movement similar to that of *Amoeba.*

Amoeboid
Refers to movement similar to that of *Amoeba.*

Amorphous
Refers to structures in which differentiation is not apparent, shapeless.

Amphibia
A class of vertebrates (q. v.) comprising frogs, toads, salamanders, newts, and related animals, most of which spend part of the life-cycle in water.

Amphibious
Refers to organisms that can live in water or on land.

Amphicarpous
Refers to plants with two kinds of fruit.

Amphichromatism
The appearance on a plant of flowers with different colors in different seasons.

Amphicryptophyte
A marsh plant with amphibious vegetative parts.

Amphidiploid
See *Amphiploid.*

Amphigean
An organism that is native to both the old and new worlds.

Amphigenesis
The union of gametes to form a *Zygote* (q. v.).

Amphigenetic

Refers to an organism or a generation that produces zygotes.

Amphimixis

Sexual reproduction in contrast to *apomixis* (q. v.). It includes *Allogamy* (q. v.), *Autogamy* (q. v.), or a mixture of these two.

Amphipodous

Refers to an animal having feet for walking and feet for swimming.

Amphiphyte

A plant growing in the border zone of wet land and water, with amphibious characteristics.

Amphiploid (Amphidiploid)

A kind of *Polyploid* (q. v.) in which there are two sets of chromosomes, each set derived from a parent in different species.

Amphirhinal

Refers to an organism with two noses.

Amphoteric

Refers to the capacity of a substance to react either as a base or an acid.

Amplectant

Refers to clasping or twining for support, e.g., tendril.

Amplexicaul

Refers to an organ clasping or growing around a stem such as the base of a leaf.

Amplitude (Ecological)

The range of an environmental condition or complex of

conditions in which an organism can exist or in which a process occurs. cf. *Tolerance.*

Anabiont

A perennial plant that produces flowers and fruits many times.

Anabiosis

Revival of an organism after apparent death, as for example by dessication.

Anabolism

The synthesis of complex organic substances from simple materials in organisms. cf. *Catabolism, Metabolism.*

Anadromous

Refers to animals having *Anadromy.*

Anadromy

The behaviour of animals such as eels and salmon which live in the sea and migrate into fresh water to breed.

Anaerobe

An organism living in the absence of free oxygen. cf. *Aerobe.*

Anaerobic

Refers to life or activity in the absence of free oxygen.

Anaerobiosis

The existence of life under anaerobic conditions.

Anaerophytobionts

The anaerobic flora of the soil.

Analogous

Refers to an organ of one organism that corresponds in function to an organ of another animal or plant but which is not *Homologous* (q. v.), e.g. petioles of clematis and leaflets of peas as twining structures, wings of birds and moths.

Anamniote

An animal lacking an embryonic membrane or amnion (q. v.), e.g. frogs, fishes. cf. *Amniote.*

Anandrous

Refers to flowers lacking stamens.

Anatomy

The branch of biology that deals with the structure of plants or animals.

Anchor-ice

Ice that has formed on the bottom of a stream.

Andean

Refers to the Andes Mountains of South America.

Androconia

Modified scales on wings of *Lepidopterons* (q. v.), producing a sexually attractive odor.

Androdioecious

Refers to the presence of flowers with only stamens and flowers with both stamens and pistils on separate plants. cf. *Andromonoecious.*

Androecium

The stamens of a flower, collectively.

Androgen

A substance causing the formation or maintenance of male sexual characteristics in certain animals.

Androgynous

Refers to flower clusters in which the staminate flowers are attached above the pistillate flowers, e.g. certain sedges.

Andromonoecious

Refers to the presence of flowers with only stamens and flowers with both stamens and pistils on the same plant. cf. *Androdioecious.*

Androphile

A plant that grows in the vicinity of man and his activities.

Anemochore

A species of which the seeds, spores, or other parts capable of reproducing offspring are dispersed by wind, e.g. dandelion. cf. *Diaspore, Disseminule.*

Anemogram

A continuous record of wind velocity made by an anemograph.

Anemograph

A self-recording instrument for recording the velocity of the wind.

Anemometer

An instrument that measures the velocity of the wind.

Anemophilous

Refers to plants in which the pollen or other spores are scattered almost exclusively by wind. e.g., willows. cf. *Entomophilous.*

Anemoplankton

Plankton (q. v.) that are transported by wind.

Anemotaxis

The reaction to wind by the movement of a free organism.

Anemotropism

The reaction to wind by the movement of an attached organism.

Aneroid Barometer

An instrument that registers atmospheric pressure in such a way that altitude may be calculated.

Aneuploid

Refers to the presence of an irregular number of chromosomes, fewer or greater than the multiple of the *Haploid* (q. v.) number. cf. *Euploid.*

Angiosperm

The subdivision of *Spermatophytes* (seed-plants) in which seeds are produced within the ovary, includes *Monocotyledons* and *Dicotyledons.*

Angle of Repose

The maximum slope on which soil or loose rock remains stable. syn. *Critical slope.*

Anhydrous

Refers to a substance that does not contain water, e.g. anhydrous ammonia.

Animal Unit

A measure of converting kinds of livestock to a common standard in relation to forage resources on the equivalent of a mature cow (live weight of about 1000 pounds). One animal unit in western range country equals about one head of cattle, one horse, one mule, five sheep, five swine, or five goats.

Animal Unit Month

A measure of forage or feed requirement to maintain one animal unit for 30 days.

Animation, Suspended

Animals of simple organization such as some nematodes, snails, and rotifers which can endure long periods of drying in an inactive condition.

Anion

A negatively charged ion, e.g., chlorine. cf. *Cation.*

Anisophylly

The presence of two kinds of leaves on one plant as in Selaginella and some junipers.

Annelid

An animal belonging to the phylum Annelida such as earthworms, leeches, and marine worms; showing segmentation.

Annual Heat Budget

See *Heat budget.*

Annual Plant

A plant which completes its life-cycle and dies in one year or less; a winter annual starts growth in late summer or fall and completes its life-cycle the following spring or early summer, a summer annual begins growth in the spring or early summer and completes its life-cycle before the following winter. cf. *Biennial, Perennial.*

Annual Production

The amount of substance formed in a year by an organism or a group of organisms.

Annual Ring

The layer of wood (*Xylem*) added each year to stems and roots of woody plants, which indicate the age of the plant. Occasionally more than one layer may be formed in one year. See *Growth layer.*

Annual Seasons

The chief climatic periods of the year; *Vernal, Estival, Autumnal (Serotinal),* and *Hibernal* (q. v.).

Annual Succession

The successive occurrence of plants or animals in an area, or their activities, during the year, such as summer-flowering plants replacing the spring-flowering ones; or the various reproductive stages in animals.

Annual Turnover

The total quantity of living organisms (Biomass) (q. v.) produced in one year in an area.

Annuation

The variation from year to year in abundance or behaviour of organisms caused often by differences in environmental conditions, especially precipitation and temperature.

Annulus

A ring-like structure characteristic of certain plant parts such as the stalk of mushrooms or on the spore-case of ferns. In certain animals, notably fishes, a ring, arrangement of rings, or other markings formed once a year and used in the determination of age and rate of growth.

Anoestrum

The period in animals when sexual desire or breeding is absent.

Anoxia

The condition of oxygen deficiency as in the tissues of an organism.

Antagonism

The depressive effect of one organism upon another one such as certain grasses upon the growth of alfalfa, or the excretion of antibiotic substances such as penicillin by a mold.

Antarctogaea

The Australian zoogeographical region, excluding New Zealand and Polynesia.

Antecedent Moisture

The degree of wetness of the soil at the beginning of a run-off period.

Anther

The pollen-producing part of the stamen in a flower.

Antheridium

The organ in which sperms are produced, found in

several groups of plants such as the algae, fungi, mosses, and ferns.

Anthesis

The period when a flower is fully expanded or when fertilization occurs.

Anthocyanin

Water-soluble pigments, usually red, blue, or violet in the cell-sap of leaves, stems, flowers, and fruits of plants.

Anthropic (Anthropeic)

Refers to the influence of man in contrast to natural influences, e.g. a fertilized soil under crop rotation.

Anthropochore

A species that is regularly disseminated by man, e.g. weeds, crop plants. cf. *Anemochore.*

Anthropogenous

Refers to influences caused by man, e.g. cultivation.

Anthropomorphism

The mode of thought or expression which attributes characteristics of man to non-human objects.

Anthropophilous

Refers to plants which grow in proximity to man such as weeds in cultivated lands or on paths.

Antibiosis

The interaction between organisms produced by an *Antibiotic* (q. v.).

Antibiotic

A substance produced by organisms, especially bacteria and fungi, which passes into the surrounding medium and is toxic to other organisms, e.g., penicillin from the mold *Penicillium notatum* destroys many kinds of bacteria.

Antibody

A substance such as *Antitoxin* produced in an animal when foreign material (*Antigen*) is introduced into the body. The antibody counteracts the effect of the *antigen*.

Anticline

A geological structure or arch formed by strata from opposite sides dipping upward toward a common line. cf. *Syncline*.

Anticryptic

Refers to coloration which facilitates aggressive action of an animal.

Anticyclone

A mass of air of high atmospheric pressure compared to adjacent areas, in which the circulation of the air is clockwise in the northern hemisphere, anti-clockwise in the southern hemisphere.

Antigen

Parasites, substances produced by them, enzymes, toxins, or proteins which cause the formation of *antibodies* in the body of an animal.

Antitoxin

An *Antibody* (q. v.) that counteracts a *Toxin* (q. v.), e.g., antitoxin serum used in treatment of diphtheria.

Apatetic

The coloration of an animal that causes it to resemble physical features of the habitat.

Aperiodicity

The irregular occurrence of phenomena. cf. *Periodicity*.

Apetalous

Refers to flowers that lack petals.

Aphis

A plant louse, an insect in the family *Aphides,* feeds by sucking juices of plants, e.g., green-peach aphid (*Myzus persicae*) which transmits more than 50 different plant viruses.

Aphotic Zone

The deeper portions of bodies of water to which daylight does not penetrate with sufficient intensity to influence organisms. cf. *Disphotic, Euphotic, Photic* zones.

Aphototropism

The response of an organism by which it turns away from the source of light. cf. *Phototropic.*

Aphyllous

Refers to plants lacking leaves.

Aphytal Zone

The part of a lake floor that lacks plants, includes the sublittoral and profundal zones.

Apocarpous

Refers to flowers in which the carpels (q. v.) are not joined together, e.g. buttercup flowers.

Apogamy

The reproduction of an organism without *Fertilization* (q. v.). cf. *Apomixis, Parthenocarpy.*

Apogeotropism

The response of an organism by turning away from the earth. cf. *Geotropism.*

Apomict Population

A population of organisms produced asexually.

Apomixis

Asexual reproduction of organisms in contrast to *Amphimixis* (q. v.). It includes *Vegetative propagation* (q. v.) and reproduction resembling sexual reproduction but in which the egg and sperm do not fuse. cf. *Apogamy, Parthenogenesis.*

Aposematic

Refers to organisms that possess coloration associated with harmful or distasteful contents, and therefore such organism may be avoided by predators.

Appetitive Behaviour

The reaction of an animal by which it becomes located or in a suitable condition to satisfy its needs.

Apron

A layer of material such as concrete or timber to protect a surface from erosion, e.g., pavement below a spillway.

Apterous

Refers to organisms or organs lacking wings, cf. *Alar.*

Aqueduct

A conduit for water, e.g., canal, pipe, tunnel, or a combination.

Aquiculture

The use of artificial means to increase the production of fish, oysters, crabs, etc., in fresh or salt waters.

Aquifer (Aquafer)

A porous soil or geological formation lying between impermeable strata in which water may move for long distances, yields ground water to springs and wells.

Aquiherbosa

Communities of herbs occurring in ponds and swamps.

Aquiprata

Communities of plants where ground water is an important factor, e.g., wet meadows.

Arable (land)

Land suitable for cultivation by plowing or tillage, does not require clearing or other modification.

Arachnid

An animal in the class Arachnida which includes spiders, mites, ticks, scorpions, and king-crabs.

Arachnoid

Refers to an *Arachnid,* particularly spiders or spider-webs; cobwebby.

Archaeophyte

A weed introduced in prehistoric time into cultivated ground.

Archegonium

The organ producing the egg in many groups of plants such as mosses, ferns, and most *Gymnosperms.*

Archibenthal

Refers to the *Archibenthic zone.*

Archibenthic Zone

The layers of the ocean between depths of about 200 feet and 3300 feet, the upper part of the *Abyssal zone* (q. v.).

Arctic

Refers to the regions in high latitudes from which tree growth is usually absent because of the shortness of the growing season and other unfavorable environmental conditions; may also be used as a noun for the region.

Arctic-alpine

Refers to the arctic and alpine regions jointly.

Arctic Life Zone

One of *Merriam's life zones* (q. v.), the portion of the *Boreal life zone* (q. v.) north of the limits of tree growth, southern limit marked by a normal mean temperature of 50°F. during the six hottest weeks of summer.

Arctogea

The faunal realm which includes the *Ethiopian, Oriental, Palearctic,* and *Nearctic* regions (q. v.). syn. *Megagea.*

Area

The total territory or range in which a *Taxon* or *Community* occurs. cf. *Basal area, Coverage.*

Areg

A sand desert.

Arenaceous

Refers to a sandy substratum.

Arenicolous

Refers to organisms inhabiting sandy substrata.

Areography

The study that deals particularly with area.

Areole

A small area of a leaf surrounded by intersecting veins.

Argillaceous

Refers to clayey material.

Arid

Refers to regions or climates which lack sufficient moisture for crop production without irrigation; precipitation 10 inches or less in cool regions, up to 15 or 20 inches in tropical regions. cf. *Semiarid.*

Aridity

The condition of dryness. See *Arid.*

Arid Transition Life Zone

The western part of Merriam's *Transition zone,* lying west of the 20-inch annual precipitation line.

Aril

A special covering on a seed, arising from base of the ovule or the stalk, sometimes pulpy or brightly colored as in the bittersweet.

Arrhythmic

Refers to the activity of an organism which occurs during both day and night. cf. *Diurnal, Nocturnal.*

Arroyo

A stream channel or gully in an arid country, usually with steep banks, dry much of the time.

Arthropod

An animal in the phylum Arthropoda such as insects, crabs, spiders, centipedes.

Artifact

(1) A substance or appearance of a specimen of an organism or preparation of part of an organism which is not present in the living tissue. (2) Something made by man especially primitive man.

Artificial Selection

Selection by man of plants or animals which possess desired qualities for reproduction and for the improvement of the properties of the organisms. cf. *Natural selection.*

Artificial Stocking

The introduction of animals from another region, or the artificial propagation of animals, into an area, e.g., stocking streams with fish or introducing quail into an area where they are scarce or lacking.

Asexual Reproduction

Reproduction of organisms without the fusion of gametes. cf. *Apomixis.*

Aspect

(1) One of the seasonal appearances of vegetation. See *Aspection.* (2) The direction toward which a slope faces.

Aspection

The variability in the appearance of vegetation or of its constituent parts such as blooming, fruiting, foliation, and defoliation during the various seasons of the year. Chief seasons are *Prevernal, Vernal, Estival, Serotinal,* and *Hibernal* (q. v.).

Assimilation

The synthesis of protoplasm and other complex substances by organisms.

Association

A term with a number of usages, some of which are defined. (1) An actual or *Concrete community, Stand,* or group of organisms characterized by a definite floristic composition, presenting uniformity in physiognomy and structure, and growing under uniform habitat conditions. (2) In an abstract sense, a group of concrete communities or stands that are classified together because they meet certain standards of similarity. See *Association type, Index of similarity.* (3) In the Clements' sense, a *Climatic climax* (q. v.) unit that includes all of the successional stages preceding or associated with it. Plant association and animal association emphasize populations of plants or animals respectively within an area.

Association, Coefficient of

A measure of the frequency of occurrence together of two species not due to chance, calculated by dividing the number of samples in which both occur together by the number of samples in which it is expected they would occur by chance alone.

Association Complex

A group of associations which occupies a definitely circumscribed area.

Association Fragment

A stand or group of plants that lacks some of the characteristics of its community type.

Association, Index of

A measure of the occurrence together of one species with another, calculated by dividing the number of samples in which one species occurs by the number of samples in which both occur.

Association, Interspecific

The occurrence together of two or more species, e.g., a parasite on a host, a grass and a legume mutually benefiting by growing in proximity to each other.

Association Segregate

A *Climax* community which has become differentiated out of a mixed or undifferentiated vegetation under the influence of climatic change, e.g., the beech-maple association arising as a segregate from the mixed deciduous forest in southeastern United States.

Association Table

A listing of species occurring in several stands of an *Association* or *Community-type,* and including data on such characteristics as abundance, cover, vitality, etc. syn. *Synthesis table.* cf. *Stand table.*

Association-type

A group of similar associations.

Associes

In the Clements' usage a temporary, developmental community. cf. *Association* (3).

Assortative Breeding (mating)

Pairing of male and female organisms that involves more than chance so that mating of similar parents is favored.

Asymptotic Population

The maximum size reached by a population under prevailing environmental conditions, no matter how long reproduction continues.

Atavism

The appearance in an organism of an ancestral character after a period of several generations.

Atlantic Period

See *Sub-boreal* period.

Atmometer

Any instrument for measuring evaporation such as a porous porcelain sphere or open pan of water.

Atoll

A body of water, lagoon, surrounded by a coral reef in the ocean.

ATP

Adenosine triphosphate, an energy-rich phosphate compound in organisms, important in the transfer of energy.

Atrophy

Reduction in size or contents of an organ, tissue, or cell.

Audiogenic Seizure

Convulsion in an animal caused by a high-pitched noise.

Aufnahme (G)

See *Sample area, Quadrat.*

Aufwuchs (G)

See *Periphyton.*

Aureomycin

The *Antibiotic* formed by the mold *Streptomyces aureofaciens.*

Auroral

Refers to the dawn or the morning *Crepuscular* period.

Austral Life-zone

One of *Merriam's life-zones* comprising most of non-montane United States and Mexico, bounded on the north by a growing season of accumulation of 10,000°F. above 43°F. and on the south by 26,000°F., with a mean daily temperature of 64.4°F. during the six hottest weeks at the northern limit.

Australian Region

The faunal region comprising Australia and New Guinea, with Tasmania and smaller islands, in the realm *Notogea* (q. v.).

Austroriparian Life-zone

A subdivision of the *Lower Austral life-zone,* east of the 100th meridian.

Autarkean Society

A simple, independent, economic human society consisting of nomadic or sparsely distributed individuals. cf. *Allelarkean society.*

Autecology

The study of the individual, or members of a species considered collectively, in relation to environmental conditions. cf. *Ecology, Synecology.*

Autochore

A species in which the action of the parent plant is the chief force of dissemination, e.g., the mechanical projection of seeds in vetch. cf. *Anemochore, Diaspore.*

Autochthonous

Refers to local origin, e.g., an *Indigenous* species, deposits produced within a lake. cf. *Index species, Allochthonous.*

Autoecious

Refers to parasites which pass all stages of their life-cycle within or on the same host, e.g., certain rust fungi. cf. *Heteroecious.*

Autogamy

(1) Self- or close-pollination leading to self-fertilization in plants. cf. *Geitonogamy, Xenogamy, Cleistogamous.* (2) Division of the nucleus into two parts followed by the union of these parts in the same cell, occurs in some diatoms and protozoa.

Autogenic Succession

A successional series in which one stage modifies the habitat in such a way that it is replaced by another stage, e.g., deciduous forest replacing a pine forest. cf. *Allogenic succession, Succession.*

Autolysis

The dissolution or digestion of an organism or parts thereof by its own enzymes.

Autonomic

Refers to processes or activities arising from internal causes, spontaneous, self-governing, e.g. *Mutation* (q. v.).

Autonomous

Refers to organisms, especially plants with chlorophyll, that are *Autotrophic* (q. v.). Also used as syn. of *Autonomic.*

Autophyte

An *Autotrophic plant.*

Autopolyploid

A *Polyploid* (q. v.) in which three or more sets of like

(homologous) chromosomes have been derived from the same species. cf. *Amphiploid, Allopolyploid.*

Autotomy

Loss of a part of the body of an organism by self-amputation followed usually by regeneration of the part, as in certain arthropods and lizards.

Autotrophic

(1) Refers to organisms which are capable of producing organic substances from inorganic materials by means of energy received from outside of the organisms, e.g., plants with chlorophyll and certain bacteria. cf. *Parasitic, Saprophytic, Heterotrophic, Holophytic.* (2) Refers to a pond or lake that is restricted in its supply of organic material to that produced within its own confines.

Autumnal

Refers to the fall season or *Aspect* (q. v.).

Auximone

A non-essential organic substance such as an extract secured from dung, of unknown chemical composition, which stimulates the growth of certain plants such as duckweeds.

Auxin

A substance, natural or synthetic, that controls the growth of plants. cf. *Hormone.*

Available Nutrient in Soil

The part of the supply of nutrient materials such as phosphates in the soil that can be absorbed by plants at rates and amounts required for growth.

Available Water (in soil)

The part of the water in the soil that plants can absorb.

Available Water-holding Capacity (of soil)

The amount of water available in the soil between the

amount held at *Field capacity* (q. v.) and the amount at the *Permanent Wilting percentage* (q. v.).

Avalanche Cone

Materials such as rocks, snow, ice, trees, deposited at the base of the path of an avalanche.

Avalanche Wind

A wind, often destructive over a distance, produced by an avalanche.

Average Distance

The distance between plants determined by dividing the square root of an area by the density of each species within the area.

Avitaminosis

An unhealthy condition or disease caused by a deficiency of vitamins.

Awn

A bristle-like structure attached to plant parts such as on *Floret* parts of grasses.

Azonal (soil)

A soil without a well-developed profile, occurring in any soil zone, consisting largely of *Parent material* (q. v.), e.g., recently deposited *Alluvium,* dune sand.

B

Back-crossing
The mating or crossing of a hybrid with either one of its parents or parental stocks.

Back-fire
A fire started purposely ahead of an advancing fire to remove inflammable material and thus control the main fire.

Backflash
The movement of a poison through root grafts from trees that have been treated with poison to non-treated trees.

Bacteriophage
A virus which destroys bacteria.

Bacteriorhizae
The nodules on the roots of most legumes and on some other plants such as alders which contain bacteria that can use atmospheric nitrogen in synthesizing organic compounds.

Badlands
Areas of rough, irregular, eroded land on which most of the surface is occupied by ridges, gullies, and deep channels, with sparse vegetation.

Baermann Funnel

A modification of the *Berlese funnel* (q. v.) for forcing nematodes out of soil or debris. The funnel is filled with warm water which forces the nematodes into a vessel below.

Baffle-pier

An obstruction placed in the path of water moving at a high velocity, e.g., a pier on the *Apron* of an over-flow dam.

Baffles

A set of vanes, guides, or similar devices placed in a conduit to check eddy currents below them, and provide a more uniform distribution of velocities.

Bag Limit

The maximum number of individuals of a species that a hunter may take legally.

Baguio

A tropical cyclone or *Typhoon,* a term used in the Philippine Islands.

Bajada

Outwash slopes with long straight longitudinal profiles, occur in southwestern United States.

Balance of Nature (Ecological Balance)

The state in an *Ecosystem* (q. v.) when the interrelationships of organisms to one another and to their environment are harmonious or integrated to a considerable degree, e.g., a climax forest. This balance may be upset in many ways such as by a drastic change in environmental conditions (erosion followed by death of many plants) or by a great increase in numbers of a certain organism (grasshoppers in grasslands). See *Dynamic equilibrium.*

Balanoid

Refers to barnacles.

Bald

A treeless area in a forest vegetation, especially in the southern Appalachians, occupied by grasses or shrubs usually.

Banados

Shallow swamps in Paraguay.

Bancroft's Law

A generalization that organisms and communities tend to come into a state of *Dynamic equilibrium* (q. v.) with their environment.

Band

A general term for a social group of two or more mobile animals of the same species, e.g., a herd of deer, pack of wolves.

Bank Storage

Water absorbed by the bed and banks of a stream and returned in whole or in part after the *Ground-water* level falls.

Bar

(1) A deposit of sand or rock particles forming a ridge along the coast, usually at the mouth of a stream or across a bay. (2) A unit of atmospheric pressure equivalent to 29.53 inches (750.1 mm) of mercury at 32°F. in latitude 45°.

Barachore

A species in which the seed, fruit, or other *Propagule* is disseminated largely by its own weight, e.g., walnut fruit. cf. *Autochore, Diaspore.*

Barchan

A distinctive isolated sand dune which is crescent-shaped with the ends projecting leeward, common in Turkestan.

Bark

A general term for all the tissues outside of the cambium in stems of trees; outer part may be dead, inner part is living.

Barogram

The continuous record made by a self-registering *Barometer.*

Barograph

A self-registering barometer.

Barometer

An instrument for measuring the atmospheric pressure. cf. *Aneroid barometer.*

Barotaxis

Response by locomotion of an organism or part of an organism in response to a barometric stimulus.

Barotropism

Reaction by growth curvature of a plant or a sedentary animal in response to a barometric stimulus.

Barrens

An area in which vegetation is absent or poorly developed.

Barrier

(1) A topographic feature or a physical or biological condition that restricts or prevents migration of organisms or prevents establishment of organisms that have migrated. (2) A condition that prevents or appreciably reduces cross-breeding of organisms.

Barrier Beach

See *Beach.*

Basal Area

(1) The area of the cross section of a tree at height of 4.5 feet above the ground, usually expressed as the summation of the basal area of the trees in a forest in square feet per acre. (2) The surface of the soil actually covered or occupied by a plant, especially the basal part, as compared to the full spread of the herbage; in grassland ecology often

measured at one inch above the ground surface. syn. *Basal cover, Ground cover.* cf. Cover.

Basal Cover

See *Basal area* (2).

Basal Metabolism

The rate of physiological processes in an organism when it is carrying on a minimum of its life processes such as respiration in order to remain alive.

Base Exchange Capacity

A measure of the absorptive capacity of a soil for bases, or exchangeable cations. A soil with a high base exchange capacity will retain more plant nutrients and is less subject to leaching than one with a low exchange capacity.

Base Flow

Stream flow originating from subterranean sources in contrast to flow from surface run-off.

Base Level

The lowest level to which a land surface can be reduced by streams; the permanent base level is the level of the sea.

Base Number

The *Haploid* (q. v.) number of chromosomes (as found in sperms or eggs, gametes) in those species with the lowest number in a *Polyploid* series, or sometimes postulated for a species that is extinct or unknown.

Base Saturation

The proportion of the *Base exchange capacity* that is saturated with metallic cations.

Basin Irrigation

A method of irrigation in which a level area is surrounded by an earth ridge so that a shallow body of water may accumulate prior to infiltration.

Basin Listing

A kind of land tillage in which small dams at intervals of 15 to 25 feet are formed across furrows to form basins for collecting water after precipitation, thus retarding runoff and erosion.

Basipetal

Refers to the development of organs in plants in which the oldest are at the apex, the youngest at the base. cf. *Acropetal.*

Basophilous

Refers to organisms which possess adaptations for life in alkaline soil or in an alkaline medium, e.g. *Atriplex* spp.

Bathyal Zone

The deep part of the ocean into which light does not penetrate effectively.

Bathybic

Refers to life in the *Bathyal zone.*

Bathypelagic

Refers to deep portions of the ocean, not including the bottom.

Bathysphere

A structure consisting of a spherical chamber in which man can descend deep into the ocean to make observations.

Bayou

A marshy body of water caused by seepage, lack of drainage, or floods, tributary to a stream or lake, in flat country. A term used in the Gulf Coast region and in the lower Mississippi River basin.

Beach

The line or zone of demarcation between land and water of lakes, seas, etc. *Barrier beach;* a ridge of deposits separated from the mainland by an interval of water.

Beach Pool

(1) Barrier beach pool is a shallow lagoon formed inland from *Barrier beaches*. (2) Sand spit beach pool is a shallow lagoon generally sigmoid-shaped inland from a sand *Spit*, characteristically on the protected side of a headland.

Beaufort Scale

A series of numbers devised by Francis Beaufort in 1805 to designate approximate wind velocities ranging from 0 for a calm to 12 for a hurricane with wind velocity in excess of 75 miles per hour.

Beckmann Thermometer

A thermometer graduated to 0.01 degree and covering a scale of 6 to 7 degrees.

Bed Load

Soil, rocks, and other debris rolled along the bottom of a stream by moving water, in contrast to "silt load" which is carried in suspension.

Bedrock

The solid rock underlying soils or other surface materials.

Behaviour, Appetitive

A reaction of an animal that aids in the satisfaction of its needs.

Behaviour, Displacement

The reaction of an animal that is not pertinent to the stimulus, as when it cannot respond appropriately or when two or more incompatible drives are present.

Behaviour, Instinctive

A fixed pattern of action that is inherited.

Belt

A comparatively narrow area or strip of vegetation with distinctive characteristics from adjoining areas or vegetation. cf. *Zone*.

Belt Transect

A strip of vegetation, usually a few inches or feet wide, in which the constituent plants are recorded or mapped.

Bench Mark

A point of reference used in elevation surveys.

Bench Terrace

A shelf-like embankment of earth constructed along the contour of sloping land to control run-off and erosion. cf. *Ridge terrace.*

Benthic

Refers to the bottom of any body of water. In the ocean the benthic division is divided into the *Littoral, Sublittoral, Archibenthic, Abyssal-benthic* zones, (q. v.). cf. *Pelagic.*

Benthos

Organisms which live on or in the bottom of the ocean or bodies of fresh water, from the water's edge down to the greatest depths. cf. *Nekton.*

Bergmann's Principle (Rule)

The generalization which states that *Homoiothermal* (q. v.) animals such as birds and mammals in cold regions tend to be larger in size and have a lower ratio of body surface to body weight than related animals in warmer climates. The reverse relationship is shown by *Poikilothermous* animals, especially amphibians and reptiles. cf. *Allen's principle.*

Berlese Funnel

An apparatus in which soil or debris is placed in a funnel, heat and light applied from above as a rule, which forces mites, collembolons, etc., into a vessel below. cf. *Baermann funnel, Tullgren funnel.*

Beta Particles

High-speed electrons given off by radioactive substances.

Beta Ray

A stream of *Beta particles* with greater power of penetration in the tissues of organisms than *Alpha rays* (q. v.).

B Horizon

A master soil horizon between A and C *Horizons* (q. v.), a layer of *Illuviation* (q. v.) in which materials from overlying horizons are deposited. B_1 is transitional between A_3 and B_2, but more like B than A. B_2 is the layer of maximum illuviation especially of silicate clay materials, or of iron and organic materials, or with maximum development of blocky or prismatic structure. B_3 is transitional between B and C horizons, often absent.

Bicentric

Refers to a *Taxon* (q. v.) with two centers of dispersion or evolution.

Biennial

A plant that lives for two years, usually blooming and fruiting only in the second year and then dying, e.g. carrot.

Biflorus

A plant that flowers in both spring and autumn.

Bilateral Symmetry

An organism which can be divided so that each half is the mirror image of the other, e.g., vertebrates, snapdragon flower. syn. zygomorphy. cf. symmetry.

Binary Name

See *Binomial.*

Binom

See *Agamospecies.*

Binomial

A name of organisms consisting of two words, e.g., *Quercus alba,* the white oak; the first name is the genus, the second the species.

Bio-assay

The employment of living organisms to test the effects of a substance such as feeding rats with food containing herbicide residues.

Biocenose

See *Biocoenosis.*

Biochemical Oxygen Demand (B.O.D.)

A test for the detection and measurement of pollution in which the quantity of oxygen that has been used by oxidizable materials under standardized conditions is determined.

Biochore

A subdivision of the *Biocycle* (q. v.) It comprises a group of *Biotopes* (q. v.) which resemble one another. The principal biochores are grassland, forest, savanna, and desert. The desert biochore includes sandy desert and stony desert biotopes. cf. *Biosphere.*

Bioclimate

See *Microclimate.*

Bioclimatic Law (Hopkins)

The generalization that in temperate North America *Phenological* events generally occur at the average rate of four days to each degree of latitude, 5° of longitude, and 400 feet of altitude; later northward, eastward, or upward in spring and early summer, and earlier in late summer and autumn.

Bioclimatology

The study of the interrelations of organisms and climate.

Biocoenology

The study of communities including qualitative and quantitative analyses; the *Synecology, Synchorology, Dynamics,* and classification of communities.

Biocoenosis (Biocoenose)

The aggregate of interacting organisms living together in a particular habitat, e.g., an oyster-bed community, usually containing producer, consumer, reducer, and transformer organisms. cf. *Ecosystem, Community, Association.*

Biocoenotics

See *Biocoenology.*

Biocycle

A subdivision of the *Biosphere* (q. v.). Biocycles usually recognized are saltwater, freshwater, and land; each consisting of *Biochores* (q. v.) cf. *Biotope.*

Biodemography

The mathematical treatment of population problems.

Bioecology

The branch of biology that deals with the interrelations of organisms among themselves and with their environments, stressing the inclusion of both plants and animals. cf. *Ecology.*

Biogenesis

The principle that living organisms can originate only from other living organisms. cf. *Spontaneous generation.*

Biogenic

Refers to biological origin.

Biogeochemical Cycle

The circulation of chemical elements such as nitrogen, carbon, etc., in specific ways from the environment into organic substances in animals and plants and back again into the environment.

Biogeocenose

A concrete or actual *Ecosystem* (q. v.), e.g., a certain bog.

Biogeographic Region

See *Biome.*

Biogeography

The branch of biology that deals with the geographic distribution of plants and animals. cf. *Plant geography, Zoogeography, Chorology.*

Biointization

Preliminary treatment for seeds with chemicals to stimulate growth.

Biological Clock

The rhythmic occurrence of processes in organisms at periodically timed intervals, e.g., the ejection of spores by the fungus *Pilobolus sphaerosporus.*

Biological Control

The use of organisms or viruses to control parasites, weeds, or other pests, e.g. control of the cottony-cushion scale by the lady beetle, prickly pear cactus in Australia by the insect *Cactoblastis cactorum.*

Biological Efficiency

The ratio of the productivity of an organism, or a group of organisms, to that of its supply of energy. cf. *Productivity.*

Biological Equilibrium

See *Biotic balance.*

Biological Factor

An influence resulting from biological as distinct from physical and chemical agents, including both *Biotic factors* (q. v.) and physiologic factors such as *Hormones.*

Biological Race (Strain)

A group of organisms which differ only in their physiological or ecological behaviour from other groups in the same species.

Biological Spectrum

A tabulation by percentages of the plants of a community

or region into the life-form classes according to *Raunkiaer's classification* (q. v.).

Biologics

Biological products such as vaccines, serums, etc.

Bioluminescence

Emission of light by living organisms such as fireflies, jelly fish, etc., popularly "phosphorescence."

Biomass

The total quantity at a given time of living organisms of one or more species per unit of space (species biomass), or of all the species in a community (community biomass). cf. *Yield, Productivity, Standing crop.*

Biome

A major biotic community composed of all the plants and animals and communities, including the successional stages of an area; the communities possess certain similarities in physiognomy and in environmental conditions. Similar to *Formation* (q. v.), e.g., the North American grassland. cf. *Biotic province, Biome-type.*

Biome-type

A group of similar *Biomes,* e. g., the temperate deciduous biome-type which includes the deciduous forests of eastern North America, China and Manchuria, and Europe.

Biometry

The application of the science of statistics to the study of organisms.

Bionomics

The study of organisms in relation to each other and to the environment. cf. *Ecology.*

Biorealm

See Biome-type.

Bios

Plant and animal life.

Bioseston

The living components of the *Seston* (q. v.).

Biosocial Facilitation

See *Facilitation, social.*

Biosphere

The portion of the earth and its atmosphere that is capable of supporting life; may be subdivided into *Biocycles, Biochores, Biotopes* (q. v.).

Biosystem

See *Ecosystem.*

Biosystematics

The study of living organisms for the purpose of recognizing and differentiating biotic units and their classification into *taxa* on the basis of genetic relationships.

Biota

All of the species of plants and animals occurring within a certain area or region.

Biothermal Zones

The divisions according to Merriam of the flora and fauna of North America on the basis of temperature data.

Biotic

Refers to life, living.

Biotic Area

A general term to denote any large area that can be delimited from adjacent areas on the basis of the composition of its *Biota.*

Biotic Balance (Biological Equilibrium)

The state of more or less self-regulation of the numbers

of plants and animals in a community, brought about by interactions within and between species and by the effects of environmental conditions. cf. *Life-cycle, Balance of nature, Pyramid of numbers.*

Biotic Climax
See *Climax.*

Biotic District
According to Dice a subdivision of a *Biotic Province,* distinguished by ecologic differences of less importance than those that separate biotic provinces.

Biotic Environment
The living parts of the environment of an organism or group of organisms.

Biotic Equilibrium
See *Biotic balance, Balance of nature.*

Biotic Factor
Environmental influences caused by plants or animals such as shading by trees or trampling by animals, sometimes used to include effects of non-living organic matter. cf. *Biological factor, Coaction.*

Biotic Formation
See *Biome.*

Biotic Influence
See *Biotic factor.*

Biotic Potential
The inherent capacity of an organism to reproduce and survive, which is pitted against limiting influences of the environment. cf. *Reproductive potential, Environmental resistance.*

Biotic Pressure
The activities of an enlarging population to maintain

itself and spread, or the tendency of one or more species to extend its range. cf. *Population pressure.*

Biotic Province

A major ecologic portion of a continent, occupying a continuous geographic area, containing one or more regional communities of plants and animals, e.g., Hudsonian biotic province which occupies most of Canada and Alaska (Dice). cf. *Biome, Formation.*

Biotic Succession

See *Succession.*

Biotin

A growth-promoting or stimulatory substance (vitamin H), a member of the vitamin B complex.

Biotope

The smallest geographic unit of a *habitat,* characterized by a high degree of uniformity in the environment and in its plant and animal life, e.g., a decaying stump, a sandy beach. cf. *Biochore.*

Biotype

A group of individuals occurring in nature, all with essentially the same genetic constitution. A species usually consists of many biotypes. cf. *Ecotype.*

Bipolar Distribution

Discontinuous distribution of a *Taxon* in the northern and southern hemispheres.

Birge's Rule

A generalization which states that the *Thermocline* (q.v.) is the transition stratum in lakes in which the temperature decreases at the rate of at least 1°C. per meter of depth.

Bisect

A line transect which shows the vertical and lateral distribution of roots along the side of a trench in the soil and the above-ground parts of the plants along the line.

Bisexual

The presence of functional male and female organs in the same plant or animal. cf. *Hermaphrodite, Dioecious, Monoecious.*

Bivalent

A pair of *Homologous* chromosomes in a certain stage of cell division.

Bivoltine

Refers to organisms with two generations a year. cf. *Univoltine, Multivoltine.*

Black Alkali

Highly alkaline soil covered with a dark incrustation of carbonates of sodium or potassium. cf. *Alkali soil, Saline soil.*

Black Earth

See *Chernozem.*

Blizzard

A storm in which the cold wind, usually of high velocity, drives fine snow and often ice crystals; the visibility is greatly reduced.

Bloom

(1) A fine pale gray granular layer, often waxy, occurring commonly on the surface of plant organs such as leaves and fruits, e. g., the grape fruit. (2) A sudden appearance of brief duration of large numbers of minute organisms, usually algae, in bodies of water.

Blowout

An excavation in loose soil, usually sand, produced by wind.

Blytt-Sernander Scheme

The chronological series of floras and kinds of vegetation following the last glaciation in Scandinavia; comprising the boreal, Atlantic, sub-boreal, and sub-Atlantic stages.

Bog

An undrained or imperfectly drained area, with a vegetation complex composed of sedges, shrubs (*Ericaceous,* especially), and sphagnum mosses, typically with peat formation; often with an area of open water. Frequently used in various meanings, in the sense of *Marsh, Swamp, Moor, Fen.* cf. *Muskeg, Heath, Raised bog.*

Bog Soil

A mucky or peaty surface horizon underlaid by peat.

Bole

The unbranched trunk or stem of a tree.

Bolson

A depression lacking exterior drainage in an arid or semi-arid region, term used in southwestern United States and Mexico.

Bonitation

The state of well being of a population, as indicated by the number of individuals. cf. *Ecological bonitation.*

Border Dike

Ridges of earth constructed to hold irrigation water within certain limits in a field.

Border Irrigation

Flooding areas in fields by the use of *Border dikes.*

Border Strip

A zone or strip surrounding a sample plot, usually given the same treatment as the plot.

Bore

A tidal wave with an abrupt front often three or more feet high, advancing upstream in a narrow river estuary.

Boreal Forest

The forest consisting chiefly of conifers extending across northern North America from Newfoundland to Alaska.

Boreal Life Zone

One of Merriam's life zones including northern North America, *Boreal forest* and *Tundra* vegetation, bounded at its southern limit by growth-season accumulated temperature above 43°F. of 10,000°F. and a mean daily temperature of 64.4°F. for the six hottest weeks. Subdivided into the *Arctic, Hudsonian,* and *Canadian* life zones.

Boreal Period

The climatic period from about 7500 to 5500 B. C., characterized by warm, dry conditions; preceded by the Preboreal period (8000-7500 B. C.) with variable climate, and the Subarctic period (9000-8000 B. C.) with cold, dry climate. cf. *Sub-boreal period.*

Bottom Deposits

Organic and inorganic materials deposited beneath water and upon the original basin or channel floor.

Bottom Fauna

Animal components of the *Benthos* (q. v.).

Bottomland

See *Flood plain.*

Boulder-clay

Unstratified clay intermixed with many stones, deposited by glaciers. cf. *Till, Drift.*

Brachypterous

Refers to organisms with short wings.

Breast Height

A height of 4.5 feet (1.3 meters) above the average ground surface or above the root collar, diameters of standing trees are ordinarily measured at this height. (Abbreviation is d. b. h.).

Breccia

A rock composed of angular pieces in a matrix.

Breeding (Plants, Animals)

The application of genetics and other sciences in the systematic improvement of a *Taxon* or a population.

Breeding Potential

See *Reproductive potential.*

Broad-base Terrace

A terrace 10 to 20 inches high, 15 to 30 feet wide, with gently sloping sides, a rounded crown, and a dish-shaped channel on the upper side, built to divert run-off water along the contour.

Broadcast Seeding

Scattering seed on the surface of the soil as contrasted to seeding with a drill in rows. cf. *Drill seeding.*

Brown Forest Soils

A group of soils with dark brown surface horizons, relatively rich in humus, becoming lighter colored below, slightly acid or neutral, with moderate amount of exchangeable calcium; commonly developed under deciduous forests that are relatively rich in bases, particularly calcium.

Brown Podzolic Soils

A group of soils with thin mats of partly decayed leaves above a thin, grayish brown layer containing mineral matter and humus; overlying yellow or yellowish brown acid *B horizons;* developed under deciduous or mixed deciduous-coniferous forests in cool temperate, humid regions.

Brown Soils

A group of soils with brown surface horizon, becoming lighter in color with depth; accumulation of calcium carbonate at depth of one to three feet; developed under grassland and shrubs in temperate to cool semiarid climate.

Browse

(1) Twigs or shoots, with or without attached leaves, of shrubs, trees, or woody vines grazed by livestock. (2) To graze plant parts as in (1).

Browse-line

A line marking the height to which browsing animals have removed the *Browse* from shrubs, trees, or vines.

Brückner Cycle

The cycle of about 35 years in average length (25 to 50 years) which includes an alternation of a warm dry period and a cold damp period.

Brushland

An area characterized by shrubby vegetation.

Brush Matting

(1) A matting of branches placed on eroded land to conserve moisture and reduce erosion while trees or other vegetation is being established. (2) A matting of mesh wire and brush used to retard streambank erosion.

Brush Pasture

A pasture with a natural cover of trees and shrubs, where a large part of the forage secured by livestock comes from browsing woody plants.

Bryocole

A small animal such as a tardigrade, rotifer, and nematode which live among moss plants.

Bryophyte

A plant in the phylum Bryophyta comprising mosses, liverworts, and hornworts.

Budding

(1) A form of grafting, by inserting a bud with a small amount of tissue at its base into a slit made in the stem of the stock plant. (2) A form of asexual reproduction in which a new cell grows out from the parent cell, e.g., yeast plants.

Buffalo Wallow

A depression in grassland made by buffalo or cattle while trampling or wallowing, followed by denudation and loss of soil.

Buffering

The modification of environmental conditions by vegetation or topographic features, e.g., shading. cf. *Reaction.*

Buffer Species

A plant or animal which may provide an alternative food for another animal and thus reduce the demand for certain food items.

Buffer Strip

A strip of grassland or other erosion-resistant vegetation planted on the contour between or below cultivated strips or fields.

Buffer Zone

(1) An area or strip surrounding a study-area or other specific area in part or entirely to protect the inner area from ecological disturbance by influences from the outside. (2) A publicly owned range area adjoining a privately owned range tract, which may be used to supplement the range on the latter.

Bulbil

A small bulb or modified bulb by which the plant is propagated.

Bulk Density

The mass or weight of oven-dry (100-110°C.) soil per unit of bulk volume, including air space.

Bunch Grass

A grass which forms a tuft or bunch, many stems arising from the root-crown in a dense mass, e.g., orchard grass, bluestems.

Bunt Order

Rank of dehorned cattle in a herd, determined by aggressive behaviour.

Buoyancy Theory

An explanation of the role of morphological features that decrease the rate of sinking of plankton.

Buried Soil

One or more layers of soil which was formerly at the surface followed by covering with ash, sand, or some other form of deposition.

Burn Scar

A scar on a tree where the tissues were damaged by fire, it may be partly or entirely covered by later-formed tissues.

Bush

(1) a shrub. (2) An area covered by shrubs or forest, especially in Australia and South Africa; also used for any uncleared land.

Butte

An isolated hill with steep sides and a comparatively flat top, smaller than a *Mesa,* term used in western United States especially.

Buttress

A lateral plank-like extension near the base of some trees, gives additional support to the tree.

C

Caatinga

A type of vegetation consisting of thorn scrub in north-eastern Brazil.

Cactoid

Resembling a cactus, e.g., *Euphorbia* spp. in northern Africa.

Caenogenesis

Special adaptations of embryonic or young stages of an animal to environmental conditions, which are more recent in evolution than adaptations in the adult, e.g., mosquito larvae with special spiracular openings. cf. *Deuterogenesis.*

Caingin Clearing

An area denuded of vegetation and used for agriculture in southeast Asia.

Cairn

A pile of stones used as a landmark.

Calcareous

Refers to material containing calcium in moderate to large amounts, especially soil with calcium carbonate.

Calcareous Ooze

Partially decomposed, soft, organic material mixed with a considerable proportion of calcareous material on the bottom of some bodies of water.

Calcicole

An organism, usually a plant, growing in soil rich in calcium.

Calcification

A soil process in which the surface soil is supplied with calcium by the decomposition of plants or in which a calcareous layer is formed in the soil. cf. *Podzolization.*

Calcifuge

A plant that grows best in acid soil. See *Acidophilous.*

Calciphile

A plant that grows best in calcareous soil. See *Basophilous.*

Calciphobe

An *Acidophilous* (q. v.) plant.

Caldera

A large basin-like depression with steep sides in the top of a volcanic mountain, e.g., Crater Lake, Oregon.

Caliche

A calcareous hardpan in southwestern United States, also applied to deposits of sodium nitrate in Chile and Peru.

Calyx

The outer whorl of flower parts, made up of sepals, usually green and resembling leaves or bracts, or with other colors in some flowers.

Cambium

A layer of *Meristem* (q. v.) cells and the undifferentiated daughter cells; used preferably for vascular cambium (which gives rise to parenchyma, secondary xylem, and secondary phloem in dicotyledons and gymnosperms) and for cork cambium.

Cambrian

Refers to the oldest geological period in the Paleozoic era, about 500 million years ago.

Campo Cerrado

A vegetation type in Brazil composed of scattered trees in dense grassland.

Campos

Grasslands or savanna located south of the equatorial forests in Brazil.

Canadian Life Zone

The southern half of the coniferous forest area of the *Boreal life zone* (q. v.).

Canopy

The uppermost layer consisting of crowns of trees or shrubs in a forest or woodland.

Canopy Trees

Trees with crowns in the uppermost layer of forest or woodland.

Capability

See *Land capability*.

Capacity, Adaptive

The genetically determined range (or plasticity) of reactions of an organism which enable it to respond in different ways to a variety of conditions.

Capacity Formula

A formula used in hydraulics to calculate the capacity or discharge volume of a channel.

Capillary Porosity

The aggregate volume of small pores within the soil which retain water against the force of gravity.

Capillary Water

The portion of soil water which is held by cohesion as a continuous film around particles and in spaces; most of it is available to plants.

Caprification

The process of *Pollination* by wasps in the commercial fig plants.

Capsule

(1) A dry, dehiscent fruit consisting of several *Carpels*. (2) A sac-like tissue surrounding an organ. (3) Organs in some plants such as mosses in which spores are produced.

Carapace

A hard case or shield covering part of the body of some animals, e.g., crabs.

Carbohydrate

An organic compound consisting of carbon, hydrogen, and oxygen, such as sugars, starch, and cellulose.

Carbon Assimilation

See *Photosynthesis*.

Carbonate Zone

A layer in the soil with a concentration of carbonates, chiefly calcium carbonate, found especially in arid regions.

Carbon Cycle

The circulation of carbon from carbon dioxide in the atmosphere into sugar by photosynthesis in plants, synthesis of more complex organic compounds in plants and animals, and the return by respiration or death and decay of plant and animal tissues to carbon dioxide.

Carbon-14 Dating

The use of radioactive carbon which has an atomic mass of 14 and an approximate half-life of 5,500 years, for determining approximately the age of soils, buried materials such as wood, and other organic materials. See *Radiation, Radioactivity.*

Carboniferous

Refers to the *Pennsylvanian* (Upper Carboniferous) and the *Mississippian* (Lower Carboniferous) geological periods in the upper part of the *Paleozoic* era, about 200-260 million years ago.

Carcinogen

A substance that produces cancer.

Cardinal Points

The four chief directions of the compass; south, east, north, and west.

Carnivore

An animal in the order Carnivora such as the dog, cat, bear, and seal.

Carnivorous

Refers to carnivores or to plants such as the sundew that trap and digest insects and other small animals.

Carolinian Life Zone

One of the divisions of the Upper Austral life zone. See *Austral life zone.*

Carotene (Carotin)

An orange-yellow pigment, a hydrocarbon, which occurs commonly in plants, especially in the roots of carrots, a precursor of vitamin A.

Carpel

The part of the flower, usually consisting of stigma, style, and ovary, the latter producing one or more ovules which develop into seeds. syn. simple pistil. Two or more carpels may be fused to form a compound pistil.

Carr

See *Fen.*

Carrying Capacity

(1) The maximum number of a wildlife species which a certain territory will support through the most critical period of the year. (2) The maximum quantity of the *Standing crop* (q. v.) which can be maintained indefinitely on an area. (3) See *Grazing capacity.*

Caruncle

(1) A protuberance, usually fleshy, near the hilum of a seed such as the castor bean. (2) A fleshy, naked outgrowth on the head and neck of certain birds, e.g., wattles of the turkey.

Caryopsis

An indehiscent, dry, one-seeded fruit in which the pericarp (ovary wall) and seed coats are united, e.g., grain of corn.

Caste

One of the kinds of specialized individuals in social insects such as termites, ants, and bees, e.g. drones, workers.

Casual Species

Species which occur rarely or without regularity in a community.

Catabolism

The aggregate of metabolic processes such as respiration and digestion by which organic compounds are changed into simpler substances. cf. *Anabolism, Metabolism.*

Catadromous

Refers to the migration of organisms, usually fish, from fresh to salt water to spawn, e.g., eel.

Catalepsy

The condition of muscular rigidity in which the body and limbs maintain the position in which they are placed.

Catalo

An animal derived by crossing cattle and buffalo.

Catarobic

Refers to an aquatic habitat in which slow decomposition of organic matter is taking place, organic substances are given off into the medium, and much oxygen is used but not enough to prohibit the occurrence of aerobic organisms. cf. *Oligosaprobic, Polysaprobic, Mesosaprobic.*

Catch Crop

A crop grown incidentally to the main crop of a farm and usually occupying the land for a short period; or a crop grown to replace a main crop which has failed.

Catchment Basin

A unit watershed, an area from which all the drainage water passes into one stream or other body of water.

Catena

A group of soils within a specific soil zone, formed from similar parent materials but with unlike characteristics because of differences in relief and drainage.

Cation

An ion carrying a positive chaıge of electricity such as calcium, sodium, and hydrogen. cf. *Anion.*

Cation Exchange

The exchange of cations held by soil absorbing materials such as calcium replacing sodium when calcium sulfate is added to a sodium-rich soil.

Cation-exchange Capacity

A measure of the total quantity of exchangeable cations that a soil can hold; preferable to base-exchange capacity.

Catkin

See *Ament.*

Caudal

Refers to the tail of an organism.

Caudate

Refers to the possession of a tail by an organism.

Caulescent

Refers to a plant which has a readily perceived stem above ground.

Cauliflorous

Refers to a woody plant that produces an *Inflorescence* directly from the trunk or one of the chief branches, e.g., the fig tree.

Cauline

Refers to the stem of a plant.

Cavernicole

An organism that lives in a cave.

Cecidium

A gall produced by an insect or a fungus on a plant.

Cell Sap

The solution in water of organic and inorganic substances in the *Vacuole* of a plant cell.

Cellulose

The principal component of cell walls of plants, a complex *Carbohydrate.*

Cenospecies

A group of species in nature distinguished only by factors external to the organisms, or the various forms of a species under domestication, a "superspecies."

Cenozoic

Refers to the geological era extending from 40 million years or more ago to the present era which began about one million years ago.

Center of Dispersal

The area from which a *Taxon* has spread or is spreading.

Center of Origin

The area in which a *Taxon* originated and from which it has spread.

Cephalic Index

A measure of the conformation of the human head, the breadth in percentage of the length (front to back).

Cereal

A plant in the grass family, Gramineae, the grains of which are used for human food, e.g., maize, wheat, and oats.

Certified Seed

Seeds that have been approved by a certifying agency as qualifying under established standards of germination, freedom from diseases and weeds, and trueness to variety.

Cespitose (Caespitose)

Refers to plants with branches, short stems that are usually covered with leaves, forming dense tufts or cushions.

Cestode

An animal in the class Cestoda, including the tapeworms.

Cetacean

A *Mammal* in the order Cetacea, including whales, dolphins, and porpoises.

Chaff

(1) A dry, thin scale found especially as bracts in flower

heads of many *Composites.* (2) The outer layers of cells of grains removed during threshing.

Chalk

A gray or white form of limestone composed mostly of the remains of small marine organisms, with a very high content of calcium carbonate.

Chamaephyte

One of the classes of *Raunkiaer's life-forms,* consisting of plants whose *Perennating* buds are located between the surface of the ground and a height of 10 inches (25 cm.).

Chaparral

Low and often dense scrub vegetation characterized by shrubs or dwarf trees with mostly evergreen and often hard leaves such as oaks and buckbrush, cf. *Maquis.*

Character

An attribute or property of an organism, functional or structural, modifiable by environmental conditions within genetically determined limits.

Character (Characteristic) Species

The species in classes 3, 4, and 5 of Braun-Blanquet's fidelity classification. Class 5 includes species occurring exclusively or almost so in a particular kind of plant community; class 4 contains species that show a strong preference for one kind of community but occurs sparingly in others; class 3 contains species that often occur in several kinds of communities but the optimum growth is found in only one kind. cf. *Fidelity.*

Characteristic Species-combination

A group of species in a community-type, which comprises the *Character species* and other species that have a *Constancy* rating above 80 per cent (varies with authors from 60 to 90 per cent).

Chart Quadrat

A chart or map of a sample area showing the location and area of each plant.

Chasmophyte

A plant growing in the crevice of rocks, e.g., saxifrages. cf. *Chomophyte.*

Check Dam

A small low dam constructed in a watercourse to decrease the velocity of stream flow and to promote the deposition of eroded material.

Chemical Stratification

A condition found in temperate lakes of the second order during the summer and winter stagnation periods in which certain horizontal strata become different chemically from adjacent ones, often with abrupt transitions.

Chemosynthesis

The kind of nutrition found in various bacteria in which energy is secured from the oxidation of inorganic materials.

Chemotaxis

Movement of an organism induced by a chemical stimulus.

Chemotrophic

Refers to an organism that obtains energy from a chemical reaction, excluding light, e.g., sulfur bacteria. cf. *Heterotrophic, Autotrophic, Phototrophic.*

Chemotropism

Growth of an organism in response to a chemical stimulus.

Chernozem

A zonal group of soils with deep, dark brown to black, fertile surface soil, rich in organic matter, grading into lighter colored soil below, and containing a calcium carbonate layer at a depth ranging from 1.5 to 4 feet. Associated with tall grassland in a temperate to cool, subhumid climate.

Chestnuts Soils

A zonal groups of soils with dark brown surface horizons grading into lighter colored soil below, and a calcium carbonate layer varying in depth from 1 to 4 feet. Associated with grassland in temperate to cool and subhumid to semiarid climates, in moister regions than *Brown soils,* drier than *Chernozem.*

Chianophile

A plant that can endure long-lasting snow-cover during winter and spring, or one that requires snow-cover in winter.

Chianophobe

A plant that cannot tolerate long-lasting snow-cover, or one that can live with little or no snow-cover during winter.

Chill-coma

The condition in animals caused by exposure to low temperature in which most of the physiological processes have been slowed down or arrested.

Chimera

An organ with genetically different tissues adjacent to each other, e.g., a green leaf with an area of white tissue.

Chinook

A warm, dry wind descending the eastern slopes of the Rocky Mountains in North America onto the adjacent plains. cf. *Foehn.* In Washington and Oregon coast country a warm, moist, southwest wind.

Chiseling

See *Subsoiling.*

Chitin

A nitrogen-containing polysaccharide forming a hard

outer layer in many *Invertebrates,* especially insects; found also in the cell walls of many fungi.

Chlorenchyma Tissue
In plants *Parenchyma* cells containing *Chloroplasts.*

Chlorinity
The chloride content of a solution, the average in sea-water is 19.3 per cent.

Chlorophyll
A mixture of chlorophyll a and chlorophyll b, the green pigments contained in *Chloroplasts* in plants, except in the blue-green algae. *Photosynthesis* is carried on in chlorophyll.

Chloroplast
The protoplasmic body or plastid in the cells of plants that contains the *Chlorophyll.*

Chlorosis
The condition of plants when chlorophyll fails to develop, the plants are yellowish white to white and poorly developed.

Chlorotic
Refers to a plant that has *Chlorosis.*

Chomophyte
A plant that grows in a fissure or crevice in rock, or on ledges where rocks have accumulated. cf. *Chasmophyte.*

Chondriosome
See *Mitochondrion.*

Chordate
An animal in the phylum Chordata, characterized by a notochord, a dorsal central nervous system, and gill slits, e.g., the *Vertebrates.*

Choripetalous
A *corolla* consisting of separate petals. syn. polypetalous.

C Horizon

In soils the unconsolidated, partly weathered rock fragments from which the upper *A* and *B Horizons* (q. v.), have developed; occasionally lacking. cf. *D horizon.*

Chorology

The study of regions or areas. cf. *Synchorology.*

Chresard

A term occasionally used for the water in soil that is available to plants for absorption.

Chromatid

One of the halves of a divided *Chromosome.*

Chromatin

Material in the nucleus and chromosomes which stains deeply with certain dyes.

Chromatophore

(1) A *Plastid* (q. v.) which contains pigment in a plant cell, e.g., *Chloroplast, Chromoplast.* (2) In animals a cell or group of cells with pigment which has the capability of changing color.

Chromogenic

Refers to the capability of an organism to produce color in a substance, e.g., certain bacteria.

Chromoplast

A plastid other than a *Chloroplast,* containing pigment, usually yellowish or red in color.

Chromosome

The threadlike or rodlike bodies bearing genes in the cells of plants and animals, formed from chromatin during the process of cell division.

Chrysalis

See *Pupa.*

Chubasco

A vortical disturbance in the vicinity of the Gulf of California, resembling *Dust whirls* on land and *Waterspouts* over water; reaches rather great heights and becomes violent enough to capsize small craft.

Chute

A high-velocity conduit for conveying water to a lower level without causing erosion because of excessive velocity and turbulence.

Chylocaulous

Refers to stems that are fleshy, e.g., cactus stem.

Chylophyllous

Refers to leaves that are fleshy, e.g., agave leaves.

Chyme

A semi-fluid substance, the partly digested food passing from the stomach into the duodenum.

Ciliate

Refers to a row of minute hairs along the margin of a structure or organ of organism.

Cilia

The hairlike, protoplasmic outgrowths on the surface of a cell.

Cinereous

Refers to color that resembles ashes.

Circle of Vegetation

All of the species and communities that are restricted, or nearly so, to a natural vegetation unit, the highest unit of floristic classification according to Braun-Blanquet.

Circulus

One of the concentric circles or ridges on a fish scale.

Circumboreal

Refers to organisms occurring in North America and Eurasia.

Circumpolar

Refers to organisms that occur in the polar regions of both hemispheres.

Cirque

A deeply eroded depression with steep slopes in areas which have been glaciated. syn. *Corrie.*

Cladode

See *Cladophyll.*

Cladophyll

A modified stem that has the appearance and the functions of a leaf, e.g., asparagus. syn. *Phylloclade.*

Clan

A group of animals that includes several interrelated families, or a group of plants arising from a common progenitor such as a group of young plants around the parent. cf. *Colony.*

Class

A unit of classification of organisms, composed of orders, e.g. *Monocotyledons, Mammals.*

Clavate

Club-shaped, larger at one end.

Clavicle

A bone in the shoulder girdle of many vertebrates.

Clay

(1) Small mineral particles of the soil, less than 0.002 mm. in diameter. (2) Soil material that contains 40 per cent or more of clay particles, less than 45 per cent of sand, and less than 40 per cent of silt.

Clay Loam

Soil material that contains 27 to 40 per cent of clay particles and 20 to 45 per cent of sand, the rest of silt.

Claypan

A layer of compact and relatively impervious clay, not cemented, but hard when dry and plastic or stiff when wet; similar to a true hardpan in that it may interfere with the movement of water or with the development of roots.

Clean Tillage

Cultivation of a field to prevent the growth of all plants except the particular kind of crop wanted.

Clear Cutting (Felling)

The felling of all merchantable trees in an area in one operation. cf. *Selective cutting.*

Clearing

(1) An area of land from which trees and shrubs have been removed. (2) One of the steps in the preparation of a tissue for microscopic observation.

Cleistogamous

Refers to *Self-pollination* (q. v.) in flowers that do not open, e.g., some violets.

Climacteric

A great change in a physiologic process, e.g., a pronounced rise in the respiration rate at about the time that some fruits such as the apple are picked, or the menopause in human beings.

Climagraph

See Climograph.

Climate

The aggregate of all atmospheric or meteorological influences, principally moisture, temperature, wind, pressure, and evaporation, which combine to characterize a region. cf.

Weather. Continental climate is the characteristic climate of land areas separated from the moderating influence of the oceans by distance or mountain barriers, marked by relatively large daily and seasonal changes in temperature. Oceanic climate is the characteristic type of climate of land areas near oceans which have a moderating influence on the range of variations in temperature.

Climatic Climax

The *Climax* (q. v.) that develops on land (moderately rolling to level) that is neither excessively nor inadequately drained in a region, so that the major environmental conditions affecting organisms are climatic, e.g., the beech-maple forest in southern Michigan. Theoretically, the ultimate phase of ecological development of communities that the climate of a given region will permit. cf. *Monoclimax, Edaphic climax, Polyclimax.*

Climatic Factors

Atmospheric or meteorological conditions which collectively make up the *Climate* (q. v.) cf. *Biotic factor, Edaphic factor, Factor ecological.*

Climatic Formation

(1) The major vegetation type in a region, e.g., the temperate climatic grassland comprising the prairie and plains grassland in the United States and Canada. cf. *Biome.* (2) A complex of communities which are geographically linked with one another because of climatic conditions, an extremely complex vegetation unit. cf. *Climatic climax.*

Climatic Region

One of the main portions of the earth's surface delimited on the basis of *Climate* such as the polar, temperate, subtropical, and tropical; each with subdivisions.

Climatograph

See *Climograph.*

Climatology

The study of *Climates* and their influences.

Climax

The kind of community capable of perpetuation under the prevailing climatic and edaphic conditions; the terminal stage of a *Sere* under the prevailing conditions. The physiographic climax is a climax determined in large measure by the nature of the topography or soil, e.g., a forest climax on a north-facing slope while grassland is the climax on the south-facing slope of the same ridge. The edaphic climax is a climax determined largely by the nature of the soil conditions, e.g., a saltgrass marsh in a poorly drained alkaline depression in grassland. A biotic climax is a climax caused by a permanent influence or combination of influences caused by one or more kinds of organisms, including man. cf. *Climatic climax, Succession.*

Climax Area

A region occupied by the same *Climax.*

Climax Community

See *Climax.*

Climax Complex

The totality of *Seres* that lead to a *Climatic climax* (q. v.), occupying a large area corresponding to a *Climatic* region.

Climax Formation

A major *Climax* occupying a large area, e.g., the deciduous forest formation. cf. *Climatic formation.*

Climax Units

The units of a *Climatic climax* (q. v.); *Association, Consociation, Society, Clan.*

Climax Vegetation

A pattern or complex of *Climax* (q. v.) stages corresponding to the pattern of environmental gradients or habitats.

Climograph

A chart in which one climatic factor such as the mean monthly temperature is plotted against another factor such as the mean monthly precipitation or the mean relative humidity.

Climosequence

A series of climatic data for different places in an area.

Cline

A gradation in genetic properties of a population along an environmental gradient.

Clinometer

An instrument for measuring the angle of a slope.

Clisere

A series of different *Climaxes* (q. v.) in a particular area resulting from changes in climate, e.g., the succession of climaxes during post-glacial time in north central United States.

Cloaca

The terminal portion of the gut into which reproductive and kidney ducts open, as for example in most *Vertebrates* such as birds, amphibians, reptiles, and many fishes.

Clod

A mass of soil produced by plowing or digging.

Clonal

Refers to a clone.

Clone

The progeny produced vegetatively, by *Apomixis*, or by *Parthenogenesis*, from a common ancestor. cf. *Ortet, Ramet.*

Closed Community

A *Community* in which the *Niches* are so well occupied by organisms that invasion by other organisms is difficult or impossible.

Closed Society

A *Society* (q. v.) in which strangers are rarely admitted, e.g., many kinds of insect societies including bees.

Cloudburst

A sudden and extremely heavy downpour of rain, especially in mountainous regions.

Cloud Forest

A forest occupying the parts of mountainous regions where cloudiness or moisture condensation occurs regularly, e.g., laurel forest in the Canary Islands.

Cloud Seeding

The placing of materials such as silver iodide in the clouds to produce precipitation.

Clutch

The aggregate of eggs or the young of birds.

Coaction

An *Interaction* (q. v.) among organisms, e.g., *Competition, Cooperation, Symbiosis.*

Coadaptation

The correlated modification of two or more mutually dependent organs or organisms, e.g., the structure of a flower and the proboscis of an insect.

Coastal Plain

A plain between the sea and higher land, usually at a low altitude.

Coccidosis
A disease in poultry, rabbits, etc., caused by certain micro-organisms (Sporozoa).

Codominant
One of two or more of the dominants in a community.

Coefficient of Association
A measure of the joint occurrence of any two species not due to chance, obtained by dividing the number of samples in which both species occur by the number of samples in which it is expected they would occur only by chance.

Coefficient of Community
The ratio of the number of species common to two communities or areas to the total number of species occurring in each of the communities. cf. *Index of similarity.*

Coefficient of Variation
The standard deviation expressed as a fraction of the mean, or as a percentage.

Coelenterate
An animal in the *Invertebrate* phylum Coelenterata, e.g., corals, sea-anemones, jellyfish.

Coelom
The body cavity in many invertebrate and vertebrate groups of animals.

Coenobiology (Cenobiology)
See *Biocoenology.*

Coenobium
A colony of organisms held together in a common substance, e.g., Volvox.

Coenocline
The sequence of natural communities in relation to

environmental gradients; the distribution of natural communities in an *Ecocline* (q. v.).

Coenocyte

A group of protoplasmic units; a structure with many nuclei and no cross walls as occurs in a number of algae and fungi.

Coenospecies

A group of *Species* in which hybridization is possible, cf. *Superspecies, Syngameon.*

Coffer-dam

A barrier constructed in a body of water so as to form an enclosure from which the water is pumped, to permit free access to the area within.

Cogonal

An artificial *Savanna* of cogongrass (*Imperata* spp.) in the Philippines.

Col

A high pass in a mountain range.

Colchicine

An alkaloid that inhibits *Mitosis,* obtained from the autumn crocus (*Colchicum autumnale*), used to produce *polyploidy* artificially.

Cold-air Drainage

The settling of cold air in low places displacing the less dense warm air, as at the mouth of a mountain canyon.

Cold Desert

Land covered with snow and ice.

Cold Front

The boundary of a mass of cold air and a mass of warm air.

Cold Hardiness (Resistance)
The capacity of an organism to tolerate low temperatures.

Coleopteron
An insect in the order Coleoptera, the beetles.

Coleoptile
The sheath surrounding the *Plumule* in the early seedling stage of plants in the grass family.

Coleorhiza
The sheath surrounding the *Radicle* in the early seedling stage of plants in the grass family.

Collembolon
A small primitive insect in the order Collembola, the springtails.

Collenchyma
Elongated, living cells with walls usually thickened mostly in the corners, common in stems of herbaceous plants.

Colloid
A substance in the colloidal state in which the dispersed particles are larger than those in a true solution, ranging from 0.001 to 0.1 micron in diameter.

Colluvium
Mixed deposits of soil material near the base of rather steep slopes, accumulations from slides, soil creep, frost action, and local wash.

Colon
The part of the large intestine excluding the rectum, of *Vertebrates*.

Colonial
Refers to organisms that form *Colonies* (q. v.).

Colonization

Occupation of an area by a group of organisms. cf. *Invasion.*

Colony

A group of individuals of one species, with a more or less permanent location, e.g., a prairie dog "town."

Columnar Structure

The arrangement of soil particles in elongated, vertical, blocky pieces with rounded tops. cf. *Soil structure.*

Combination of Species, Characteristic

See *Characteristic combination of species.*

Commensal

One of the organisms reacting in *Commensalism* (q. v.).

Commensalism

The living together of two or more organisms with benefit usually to one and with injury to none. cf. *Symbiosis, Coaction.*

Commensurability

A measure of the extent to which ranches should share in grazing privileges on nearby public land or cooperatively controlled range, as determined by the forage resources of the privately controlled property of the ranches involved.

Community

A group of one or more populations of plants and animals in a common spatial arrangement; an ecological unit used in a broad sense to include groups of various sizes and degrees of integration. cf. *Association, Biocoenosis,* Concrete community, See *Stand.* An Abstract community or Community-type is an assemblage of stands, e.g., the oak-hickory community-type. cf. *Association.* A *Microcommunity* is a community or stand occupying a small area such as an area of mosses between clumps of grass and a *Microcommunity-type* consists of an assemblage of microstands. cf. *Closed community.*

Community Complex

A mixture of concrete communities or *Stands,* including transitional stands, e.g., a sand-dune complex.

Community Dynamics

The aggregate of changes that take place within and between communities. cf. *Succession, Syngenetics, Fluctuation.*

Community Mosaic

The arrangement of two or more microstands making up the plant and animal life of an area, such as the different kinds of vegetation on the mounds and in the depressions in a marsh or bog.

Community Regulation

See *Homeostasis.*

Community-type

(1) See *Abstract community.* (2) A group or class of similar abstract communities.

Companion (Species)

According to Braun-Blanquet's *Fidelity* classification the species of plants that are not restricted to any definite kind of vegetation unit.

Companion Crop

A crop which is grown with another crop, usually applied to a small grain crop ("nurse crop") with which forage crops are sown.

Compatibility

The capacity of two organisms to crossbreed successfully.

Compensation Intensity

The intensity of light at which the amount of oxygen

produced by *Photosynthesis* of a plant equals the oxygen absorbed in *Respiration.*

Compensation Level (Point)
The depth in a body of water at which the *Compensation intensity* of a given plant occurs.

Competition
The condition that exists when the requirements of one or more of the organisms living in a community cannot be obtained from the available supply of resources. cf. *Exploitation, Interference.*

Competitive Exclusion Principle
A generalization that states that as a result of competition two similar species rarely if ever occupy the same ecological *Niche.* Also termed Gause's principle, Grinnell's axiom.

Competitor
An organism competing with one or more other organisms.

Complementary Genes
Two or more genes that by their joint action produce a character.

Complete Flower
A flower that has all of the usual parts; sepals, petals, stamens, and one or more pistils.

Complex
See *Community complex.*

Complex Gradient
A gradient comprising a mixture or a combination of environmental conditions. cf. *Ecocline.*

Composite
A plant in the family Compositae, e.g., aster, sunflower.

Compost

A pile of decomposing organic matter of plant or animal origin in which soil or other amendments such as lime, nitrogen, and phosphorus may be mixed.

Concealing Coloration

Color of plumage, pelage, scales, scutes, skin, or other body covering which brings about some degree of conformity in the appearance of an animal with its biotic or inanimate environment.

Concentrates

Feed that has a high content of total digestible nutrients and low fiber content, e. g., grain and grain by-products.

Conditioned Reflex

See *Reflex*.

Conditioner

A substance that modifies the characteristics of a material or medium to which it is added.

Condition Factor

A numerical index, usually applied only to fishes, which represents the relationship between length and weight of the animal.

Conditioning, Environmental

The modification of the environment of one or more organisms by their activities including *Reactions* and *Coactions,* e. g., liberation of oxygen by water plants in an aquarium.

Conductivity

The total electrolytic content of natural waters, determined by measuring the electrical conductivity.

Congelifraction

The splitting of rocks caused by frost.

Congeneric

Refers to plants or animals in the same genus, e. g., *Quercus alba* and *Q. rubra*.

Conifer

Any plant in the order Coniferales, e. g., pine, spruce, fir, juniper, etc.

Coniferous

Refers to a *Conifer,* or to the order, Coniferales.

Conjunctive Symbiosis

See *Symbiosis.*

Conservation

Usage or the aggregate of practices and customs of man that permit the perpetuation and sustained yield of renewable resources and the prevention of waste of non-renewable resources.

Consociation

In the Clements usage a morphological part of a *Climax Association,* characterized by the presence of one dominant, e. g., the little bluestem and needlegrass consociations in climax tall grass or true prairie.

Consocies

In the Clements sense a morphological part of an *Associes* (q. v.), a developmental unit, characterized by the presence of a single dominant, e. g., a stand of Russian thistle in the first weed stage in secondary *Succession.*

Consortism

See *Symbiosis.*

Constance

See *Constancy.*

Constancy

The percentage of occurrence of a species in the total

number of plots, uniform in area, located in a number of stands of one kind of *Community-type* or *Abstract community*. cf. *Presence.*

Constants of a Community or Association

The species which show the highest degrees of *Constancy* (q. v.); the most usual lower limit is 80 per cent, but varies from 50 to 90 per cent, according to various schools.

Constructive (Species)

Refers to plants whose *Reactions* or *Coactions* aid in the development or persistence of a *Community.*

Consumer Organisms (Consumers)

Organisms which ingest other organisms or food particles, may be classified as primary, secondary, etc., depending upon their position in the *Food chain* (q. v.) or the *Trophic level* (q. v.). cf. *Producers.*

Consumptive Use

The quantity of water used and transpired by vegetation plus the amount lost by evaporation. syn. *Evapotranspiration.*

Contagious Dispersion

The non-random (above normal) occurrence of individuals of a species, forming aggregations. syn. over-dispersion, *Hyperdispersion.* cf. *Normal dispersion, Hypodispersion.*

Continental Bridge Hypothesis

The hypothesis that the present-day continents were once connected by isthmuses, or other areas of land.

Continental Drift Hypothesis

The hypothesis, advanced especially by Wegener, that the present-day continents were displaced horizontally from the original mass of land to their present positions.

Continental Platform

The parts of the world comprising the lower areas of continents and the *continental shelves* (q. v.).

Continental Shelf

The shallow, gently sloping portion of the seabottom bordering a continent, down to a depth of about 100 fathoms.

Continental Slope

The steeply sloping portion of the sea-bottom extending seaward from the *Continental shelf.*

Continuous Grazing

The practice of grazing the vegetation of an area without interruption throughout the season. cf. *Deferred grazing, Rotation grazing.*

Continuum (Vegetation and Animal Life)

The occurrence of populations of organisms along a gradient, forming a distribution pattern of intergrading populations.

Contour

(1) An imaginary line on the surface of the land which connects points of the same altitude. (2) A line on a map to show the location of points of the same altitude.

Contour Farming

The performance of farming operations such as plowing, seeding, and cultivating along contour lines.

Contour Furrows

Furrows located along contour lines on range or pasture land to prevent or retard runoff and permit the infiltration of water into the soil.

Contour Interval

The vertical distance between two contour lines.

Contour Strip Cropping or Farming

The growing of crops on the strips between contour lines, at right angles to the slope. Strips of grass or other plants may be grown in alternation with the cultivated crops. A conservation practice to control or eliminate runoff and erosion, and permit greater infiltration of water.

Control Factor

The chief limiting factor or condition influencing an organism, e. g., wilting of a plant caused by insufficient soil water.

Control Flume

An open conduit or artificial channel arranged for measuring the flow of water.

Controlled Burning

See *Prescribed burning.*

Convergence

The increase in similarity of different *Seres* as *Succession* proceeds from early to late stages.

Convergent Evolution

The development of similarity in characteristics of organisms that were originally more different.

Cooperation

The kind of reaction between organisms which are beneficial and non-obligatory to those participating. cf. *Disoperation, Protocooperation, Coaction.*

Copepod

An animal in the order Copepoda, minute *Crustaceans* in salt and fresh water.

Coppice

A grove in which the trees are regularly cut, new growth arising from the base.

Copraphagous
Refers to organisms that feed on dung.

Copse
See *Coppice.*

Coral Reef
A series of calcareous rocks formed chiefly by corals, partly by algae, at or near the surface in some warm parts of the sea.

Cordillera
A system of mountain ranges, e. g., the Andes Mountains in South America.

Coriaceous
Refers to structures that are leathery such as leaves.

Corm
A short, firm, enlarged, fleshy underground stem as in the crocus.

Cormophyte
A plant that has a stem and roots. cf. *Thallophyte.*

Corneous
Refers to a structure that is horny in texture.

Corolla
The whorl of parts, usually colored, of a flower, composed of petals, within the calyx.

Corraision
The process by which flowing water carrying solid material wears away underlying rock, e. g., a stream carrying gravel and sand.

Corridor
A broad, continuous land connection enduring a long time and thus permitting the extensive interchange of or-

ganisms by migration as at the present time between Asia and Europe. cf. *Filter bridge, Sweepstakes bridge.*

Coteau

A series of *Moraines,* a term used in Western United States.

Coterie

A closed social group of animals, individuals of which defend their common territory against members of other coteries.

Cotyledon

A primary leaf of the embryo in seeds, only one in the *Monocotyledons,* two in *Dicotyledons.* In many of the latter such as the bean they emerge above ground and appear as the first leaves.

Cover

(1) The plants or plant parts, living or dead, on the surface of the ground. Vegetative cover or herbage cover is composed of living plants, litter cover of dead parts of plants. cf. *Basal area.* (2) The area of ground covered by plants of one or more species.

Cover, Vegetation

The area of ground covered by the sum total of plants in an area.

Coverage

The percentage of the area of a community covered by a plant or an animal that is attached to the substratum, as seen from above. cf. *Cover.*

Cover Crop

A crop growing close to the ground for the chief purpose of protecting the soil from erosion and also for the improvement of its fertility, between periods of regular production of the main crops, or between trees and vines in orchards and vineyards.

Covert
A place of concealment for an animal, e.g., a hedge row.

Cover Type
The present vegetation on an area, a community forming the *Cover* at the present time.

Cow Month
The quantity of feed or forage required for the maintenance of a mature cow in good condition for 30 days. cf. *sheep month.*

Creche
A group of young animals after leaving their nests.

Creek
A stream that is intermediate between a river and a brook.

Creep, Soil
The slow, downward, mass movement of soil on a slope.

Crepuscular Periods
The periods of dusk before sunrise and after sunset. cf. *Auroral, Vesperal, Diurnal, Diel, Nocturnal.*

Cretaceous
The most recent geological period of the *Mesozoic* era, which began abont 135 million years ago and lasted for about 60 million years.

Critical Factor
See *Limiting factor.*

Critical Slope
See *Angle of repose.*

Cropland

Land that is used regularly for the growing of crops (except forest crops and permanent pasture). Includes orchards, cultivated summer fallow, rotation pasture, and land that is temporarily idle but customarily used for production of crops.

Crop Residue

The parts of plants, or a crop, left in the field after harvesting the desired part such as grain or fruit, e. g., *Stubble.*

Crop Rotation

The growing of different crops in recurring succession on the same piece of land.

Crop, Standing

The total amount of organic material of one or more species in a certain space at a given time; e. g., the trees in a stand that are useful for lumber or other products. cf. *Biomass.*

Cross

See *Hybrid.*

Cross-fertilization

Sexual reproduction by means of two separate organisms. cf. *Self-pollination, Cross-pollination.*

Crossing-over

The interchange of parts of *Chromatids* (q. v.) of *Homologous chromosomes* during pairing in *Meiosis.*

Cross-pollination

The transfer of *pollen* from the anther in the flower of one plant to the stigma in the flower of another plant. Syn. *Xenogamy.* cf. *Self-pollination.*

Cross-timbers

Strips of oak forest at right angles to the river systems in Oklahoma and Texas.

Crown Canopy

See *Crown cover.*

Crown Class

The trees occupying a similar layer or position in the crown cover such as the dominant crown class consisting of dominant trees in the canopy layer; co-dominant crown class, trees with less well developed crowns but in the canopy layer; intermediate crown class, trees with crowns mostly below the canopy layer but extending into it; and overtopped crown class, trees with crowns entirely below the canopy layer, crowns poorly developed, or trees are suppressed, dying, or dead.

Crown Cover

The canopy formed by the trees in a forest.

Crown Density

The percentage of the total area of land that has a complete crown cover.

Crucifer

A plant in the mustard family, Cruciferae, e. g., radish.

Cruise

A survey of forest land to locate and estimate the volume and grades of the standing timber; or an estimate secured in such a survey.

Crumb Structure

The condition of a soil that contains irregularly shaped and highly porous aggregates.

Crustacean

An *Arthropod* in the class Crustacea, e. g., crab, shrimp.

Crustal Movement

A movement of the outer solid part of the earth such as an earthquake.

Cryophyte

A plant growing on snow or ice, e. g., "red snow," an alga, *Chlamydomonas* sp.

Cryoplanation

Land erosion or reduction by intensive frost action.

Cryptogam

A plant in any of the groups; *Thallophytes, Bryophytes,* and *Pteridophytes.* cf. *Phanerogam.*

Cryptophyte

A plant in one of the *Life-form classes* of Raunkiaer in which the buds are covered with soil or water; includes *Geophytes, Helophytes,* and *Hydrophytes* (q. v.).

Cryptozoic

Refers to animals living in darkness as under stones or in caves.

Cucurbit

A plant in the gourd family, Cucurbitaceae, e. g. squash, cucumber.

Cuesta (Spanish)

A kind of ridge with a gentle slope on one side and a steep slope on the other.

Culled Forest

A *cut-over forest* from which only certain individuals or species have been removed, e. g., a forest culled for pines of a certain minimum diameter.

Culm

A stem, especially the grass stem with nodes and internodes.

Cultch

Empty shells and other kinds of material dumped into

spawning areas to provide a suitable substratum for growth of oysters.

Cultigen

A plant or a group of plants that is grown only under cultivation so far as is known, e. g., cabbage. cf. *Indigen.*

Cultivar

A strain, variety, or race which originated and is maintained under cultivation; not necessarily a species.

Culture Community

A *Community* brought about by man's activity, e. g., a seeded meadow; or a natural community greatly altered by man. cf. *Hemerocology, Secondary succession.*

Curie

The basic measure or unit of radioactivity of a substance, the disintegration of a *Radioactive Isotope* at the rate of 3.7 times 10^{10} atoms of material per second.

Cushion Plant

An herbaceous perennial plant that produces a form with a dense mass of short stems and many leaves.

Cuticle

A covering of fairly water-proof material composed of *Cutin* in higher plants, or chitin and/or protein in many animals.

Cutin

A mixture of waxlike materials forming the *Cuticle* of higher plants.

Cut-over Forest

A forest from which some or all of the merchantable trees have been removed. cf. *Culled forest.* syn. *Logged-over.*

Cybernetics

The study of kinds of communication and control systems in human beings and in machines.

Cycle (Population)

The regular or approximately regular oscillation in the abundance of a population or species. An *Intrinsic cycle* is caused by the interactions within populations of one or more species, an *Extrinsic cycle* is caused by changes in the physical or biotic environment.

Cycle of Erosion

The changes brought about by erosion from youthful to mature to old-age *Topography*.

Cyclomorphosis

The change in form of some animals in accord with the season of the year as occurs in *Cladocera*.

Cyclosis

The streaming of *Cytoplasm* in plant cells.

Cytogenetics

The combination of *Cytology* and *Genetics* in the study of variation in organisms.

Cytology

The study of cells of organisms, a branch of biology.

Cytolysis

The disintegration of a cell.

Cytoplasm

The *Protoplasm* of the cell excluding the *Nucleus*.

Cytotaxonomy

The combination of *Cytology* and *Taxonomy* in the study of classification of plants.

D

Damping-off

The sudden wilting and death of seedling plants, caused by microorganisms.

Day-degrees

The sum of degrees of heat above a threshold, such as the sum of the degrees above a daily mean of 43°F. for the growing season, or for some other period. cf. *Temperature summation.*

Day-neutral Plant

A plant that blooms when the length of day is either long or short. cf. *Photoperiodism.*

D. B. H.

See *Breast height.*

DDT

An *Insecticide* (q. v.), dichloro-diphenyl-trichlorethane.

Dealkalization

Removal by leaching or by chemical treatment of exchangeable sodium from the soil. cf. *Alkali soil.*

Decapod

An animal in the order Decapoda, class Crustacea, e. g., lobster, crab, shrimp.

Decidulignosa

Communities consisting of trees or shrubs with deciduous leaves.

Deciduous

Refers to the losing of parts of an organism such as leaves of trees or antlers of deer at certain seasons.

Decomposer Organism

An organism, usually a bacterium or a fungus, that breaks down the bodies or parts of dead plants and animals into simpler compounds.

Decreaser

A species that decreases in *Population density* or *Cover* under continued grazing. cf. *Increaser.*

Deferred Grazing

The postponement in the grazing of vegetation after growth has started until a certain stage of development has been attained in order to promote vigor of the plants. cf. *Continuous grazing, Rotation grazing.*

Deficiency Disease

A disease or malfunctioning of an organism caused by the lack or insufficiency of some food substance such as a certain vitamin or a mineral.

Deflocculation

The separation of soil aggregates containing clay into individual particles.

Degradation (Soil)

The change in a soil that occurs in leaching, e. g., a *Chernozem* into a *Podzol.*

Dehiscent

Refers to a structure that breaks open at maturity, e. g., a pea pod. cf. *Indehiscent.*

Deme

One or more populations of a *Taxon,* an interbreeding population.

Demography

The study concerned with the analysis of populations including births, deaths, age, etc.

Dendrochronology

The dating of events or historical periods by the study of growth rings of trees.

Dendrology

The study of trees.

Denitrification

The change of nitrogenous compounds by certain bacteria in which free nitrogen is formed.

Density (Population, Species)

The number of individuals in relation to the space in which they occur, refers to the closeness of individuals to one another. cf. *Population density, Abundance, Cover.*

Density-dependent Factor

An influence that is dependent upon a certain density of individuals in order to be fully effective, e. g., a limited number of prey animals for the number of predators present in an area.

Density-independent Factor

An influence that is effective without regard to the density of individuals in a population, e. g., very unfavorable weather such as a blizzard. According to Andrewartha (1954) and Birch this factor is non-existent. See *Non-reactive factor.*

Denudation

(1) The processes by which the surface of the earth is worn away, including rainfall, wind, erosion, waves, tides, frost action, heating by the sun, etc. (2) The total destruction of plant and animal life in an area by physical or biotic means.

Dependence

A relationship between organisms in which one organism receives benefit from the other, not reciprocal. cf. *Competition, Symbiosis, Coaction.*

Dependency Zone (Range)

A certain area surrounding an area of private land, within which the use of the private land may be supplemented by use of the public land.

Dependent Property (Range)

Privately owned or controlled land or water judged to have special claims for companionate use with certain public or cooperatively controlled range land.

Deposits

See *Allochthonous, Autochthonous, Terrigenous.*

Depth, Effective Soil

The depth of the soil which roots of plants can penetrate readily to obtain water and plant nutrients. cf. *Working depth.*

Desalinization

Removal of salts from a saline soil, usually by leaching.

Desert

An area of land which has an arid, hot to cool climate, with vegetation that is very sparse and usually shrubby.

Deserta

Various kinds of vegetation found in areas and on substrata that are poor in available water for plant growth such as dry deserts, salt deserts, cold deserts, strand vegetation, dune communities, and rock communities.

Desert Grassland

The extensive grassland in southwestern United States and Mexico, characterized in part by several species of grama-grass, three-awn grass, and curly mesquite.

Desert Pavement

The stony or pebbly surface of land after the fine materials have been removed by wind or water action.

Desiccation

See *Siccation.*

Desilting Area

An area occupied by vegetation such as grasses or shrubs used solely for the deposition of silt and other debris from flowing water, located above a reservoir, pond, or field which needs protection from accumulation of sediment.

Desilting Basin

See *Settling basin.*

Detention Dam

A dam built for the purpose of storing streamflow or surface runoff, and to control the release of such stored water.

Deuterogenesis

The development of adaptive characteristics in late stages of the life-cycle, e. g., wings of insects. cf. *Caenogenesis.*

Devonian

A geological period in the *Paleozoic* era, which began about 325 million years ago and lasted about 45 million years.

Dew-point

The temperature at which a certain body of air is capable of holding no additional water vapor, so that any decrease in temperature or any increase in water vapor will result in condensation of the vapor into liquid water; at this point the *Relative humidity* (q. v.) is 100 per cent and the *Saturation deficit* (q. v.) is zero.

D Horizon

The stratum in the soil below the depth of weathering, composed of undifferentiated and unconsolidated parent materials, immediately below the *C horizon. cf. A horizon, B horizon.*

Diameter Breast High

See *Breast height.*

Diapause

A period of suspended growth or development and reduced metabolism in the life-cycle of many insects, in which the organism is more resistant to unfavorable environmental conditions than in other periods.

Diaspore

A portion of a plant such as a seed, spore, bud or other part that undergoes dispersal and can give rise to a new plant. cf. *Disseminule.*

Diastrophism

Dislocation of the earth's crust such as folding, resulting in the formation of mountains, sea basins, etc.

Diatom

A one-celled, microscopic alga in the class Bacillariaceae, with siliceous walls.

Diatomaceous Earth

A deposit of the siliceous remains of diatoms.

Diatomaceous Ooze

Material consisting of siliceous remains of diatoms found in cold seas.

2, 4-Dichlorophenoxyacetic Acid (2, 4-D)
A compound used to destroy undesirable plants, applied as a dust or spray to the foliage.

Dichogamy
The maturing of stamens and pistils of a flower at different times.

Diclinous (Diclinic)
Refers to plants that have stamens and pistils in separate flowers. cf. *Monoecious, Dioecious, Monoclinous.*

Dicotyledon
A *Vascular* plant in the subclass Dicotyledoneae, class Angiospermae (flowering plants), which have seeds containing two seed-leaves or cotyledons, e. g., peas, beans. cf. *Monocotyledon.*

Diel
Refers to the 24-hour period of day and night. cf. *Diurnal, Nocturnal, Crepuscular.*

Differential Species (Differentiating Species)
A species, because of its greater *Fidelity* (q. v.) in one kind of community than in other kinds can be used in distinguishing vegetation units.

Differentiation
(1) The development of a cell, organ, or immature organism into a mature organism. (2). The development of different kinds of organisms in the course of evolution.

Digestion
The conversion of complex, usually insoluble organic substances into simpler and usually soluble compounds by enzymes.

Dihybrid (Cross)
An organism resulting from the breeding of parents that

differ in two characters such as color of flowers and length of stems. cf. *Monohybrid.*

Dimorphism

The state of organs of a plant or animal or individuals in a population occurring in two forms or colors, e. g., a plant with leaves of two forms. cf. *Polymorphism.*

Dinoflagellate

A motile organism in the class Dinophyceae of the algae; great abundance of some forms ("red tides") along coasts causes death of many fish.

Dioecious

Refers to plants with pistillate and staminate flowers in separate plants, e. g. willows. cf. *Monoecious. Diclinic.* In animals refers to unisexual organisms. cf. *Hermaphrodite.*

Diploid

Refers to the presence of chromosomes in pairs or in two sets, resulting from the union of two *Gametes,* each with a single set (*Haploid,* q. v.).

Dipterous

Refers to an insect in the order Diptera which possess two wings (except parasitic forms), e. g., housefly, mosquito.

Disclimax

An enduring *Climax* (q. v.) community altered by disturbance by man or domesticated livestock, e. g., a grassland which has replaced a deciduous forest. cf. *Plagioclimax.*

Discontinuity

The existence of a gap in the geographic distribution of a *Taxon.* cf. *Distribution, Disjunct.*

Disjunct

Refers to the absence of a connection as in the geographic distribution of a *Taxon* or a community. cf. *Discontinuity.*

Disjunction
See *Discontinuity*.

Disjunctive Symbiosis
See *Symbiosis*.

Disoperation
An interaction between organisms in which one or all are harmed, e. g. *Competition* resulting in stunted growth.

Dispersal
(1) The actual transfer or movement of *Disseminules* or organisms from one place to another. (2) The history of the movement of a group of organisms. cf. *Migration, Establishment, Spread*.

Dispersion
The pattern of *Distribution* of individuals of a *Population*, especially in regard to probability.

Dispersion (Soil)
The breaking down of soil aggregates, resulting in single grain structure; usually the more easily a soil is dispersed the more erodible it is.

Disphotic Zone
The depths in bodies of water where light is inadequate for photosynthesis in plants but adequate for animal life. cf. *Aphotic zone, Euphotic zone*.

Displacement Theory
See *Continental drift hypothesis*.

Dissemination
The processes by which organisms or their parts, especially spores, seeds, or fruits are scattered. cf. *Diaspore, Dispersal, Disseminule*.

Disseminule

A detachable part of a plant which is capable of *Dispersal* and of giving rise to a new plant, cf. *Diaspore.*

Distribution

(1) The geographic range (continuous or discontinuous) of a *Taxon* at any one time. (2) The pattern of occurrence of individuals of a taxon in an area such as *Random* or *Poisson (Normal)* distribution; non-random or above normal, *Contagious dispersion (Over-dispersion, Hyperdispersion);* and non-random, below normal, *Hypodispersion* or even-spaced.

Distribution, Center of

See *Center of dispersal.*

Diurnal

Refers to daytime in contrast to *Nocturnal.* cf. *Diel, Crepuscular.*

Divergence

The condition in which *Seres* of a similar origin become increasingly unlike as *Succession* proceeds toward the *Climax.* See *Convergence.*

Diversion Dam

A dam constructed for the purpose of diverting part or all of the water in a stream into a different course.

Diversity Index

The number of species in an area divided by the number of individuals of all these species.

Division of Labor

The specialization of parts of an organism or members of a species for carrying on certain processes, e. g., in birds the wings for flying and the legs for walking, in bees the workers and drones, in plants the various kinds of tissues for carrying on different functions.

DNA

Deoxyribonucleic acid, the chief constituent of chromosomes which apparently is the material constituting the genes.

Doldrums

The equatorial belt of calm or light variable winds, low atmospheric pressure, lying between two trade-wind belts.

Dolomite

A rock containing a high percentage of calcium and magnesium carbonates.

Domatium

A small structure on certain plants, particularly on the leaves, which forms a shelter for organisms such as insects or fungi.

Dominance Classes

The five groups of species in a classification based on *Coverage* (q. v.).

Dominance, Ecologic

The condition in communities or in vegetational strata in which one or more species, by means of their number, coverage, or size, have considerable influence or control upon the conditions of existence of associated species.

Dominance, Genetic

The influence exerted by a dominant character or *Allele* e.g., redness of petals in certain flowers is dominant over white. cf. *Recessive*.

Dominance, Social

The determination of the behaviour of one or more animals by the aggressive behaviour or otherwise of other individuals, resulting in the establishment of a social *Hierarchy*.

Dominant (Character)

See *Allele*.

Dominant (Species)

A species that manifests *Ecologic* or *Social dominance.* cf. *Secondary species.*

Donga

A gully with steep sides or a dry watercourse, a term used in South Africa.

Dormancy

The condition in an organ or in an organism where metabolic processes are relatively inactive as a result of internal causes, e. g., many kinds of seeds, overwintering stages of insects. cf. *Hibernation, Estival, Diapause.*

Double Fertilization

The process, unique in *Angiosperms,* in which one male nucleus fertilizes the egg nucleus to form the *Zygote* which develops into the *Embryo,* and the other male nucleus joins with two other nuclei in the embryo sac to form the *Endosperm,* e. g., in corn and other grasses.

Down

An area of open, treeless upland with a thin covering of soil, used mostly for sheep grazing; especially the chalk hills of southern England.

Downland

Temperate grasslands in Australia and New Zealand.

Drainage Basin

The largest natural drainage area subdivision of a continent, such as the Mississippi, Columbia, and Colorado basins. cf. *Watershed.*

Drainage Terrace

A *Graded terrace* constructed to have a relatively deep channel and a low ridge primarily for drainage of a hillside.

Draw

A natural depression or swale; a small natural drainage-way.

Drift (Geology)

Material of any kind which is deposited in one area after having been moved from another, most commonly used in reference to glacial drift, the material deposited by glacial action. Glacial drift includes *Till* (q. v.) and stratified outwash materials. cf. *Loess, Boulder-clay, Drumlin.*

Drift Barrier

An open structure constructed across a stream channel to catch driftwood, such as a wire fence.

Drift Fence

A fence for the purpose of preventing livestock from going from their regular range to another, often used in connection with natural barriers.

Drift, Genetic

Random changes in the characteristics or attributes of populations that are usually isolated, or in the frequencies of certain genes, which cannot be attributed to selection, mutation, or migration. cf. *Natural selection.*

Drift Ice

Portions of icebergs or ice-floes in the open sea outside of the areas of pack-ice.

Drill Seeding

Sowing seeds with a drill usually in rows that are less than one foot apart as in seeding grains. cf. *Broadcast seeding.*

Drip Point

The long, attenuate tip of many leaves in the *Rain forest* of the *Tropics.*

Drive

The complex of internal and external states and stimuli leading to a certain behaviour in an animal.

Driveway, Stock

A strip of land set aside for the movement of livestock from one place to another.

Drosophila

A genus of flies, order Diptera, much used in research in genetics, e. g. fruit-fly.

Drought

An extended period of dryness; usually any period of moisture deficiency that is below normal for a specific area.

Drought Resistance

The capability of an organism to survive drought with little or no injury.

Drumlin

An oval-shaped hill composed of glacial *Drift*, usually compact and not stratified, commonly with its longer axis parallel to the movement of the ice when deposition occurred.

Drupe

A fleshy fruit in which the single seed is within a stony inner cover (endocarp) which is surrounded by a fleshy layer (pericarp), e. g., plum, cherry.

Drupelet

A small drupe, e. g., the raspberry fruit is a cluster of drupelets.

Dry Farming

(1) Cultivation of land and other farming operations in semi-arid or arid regions without irrigation. (2) A system of cultivation of the land in which *Fallow* and *Mulch* are used to absorb and retain much of the precipitation that occurs.

Duckfoot

An implement with horizontally spreading V-shaped blades which provide shallow cultivation without turning over the surface soil or entirely burying crop residues.

Ductless Glands

See *Endocrine gland.*

Duff

A general term for vegetal material in forests, including the fresh litter and well decomposed organic material and humus. See *A horizon.*

Dune

A mound or ridge of sand piled up by the winds, commonly found where sand is abundant as along lake shores, sea shores, and in desert and semi-desert areas.

Dune Sand

(1) An area of sand accumulated by wind action into dunes or hummocks, usually free from vegetation or sparsely vegetated and undergoing erosion and redeposition by the wind. (2) Refers to sand that has texture size of 0.1 to 0.4 mm. in diameter which has been piled up by the wind.

Duriherbosa

Vegetation consisting of herbaceous plants whose above-ground parts die during winter, e. g., grasslands.

Durilignosa

Vegetation consisting of broad, hard-leaved *Sclerophyll* trees or shrubs, e. g. *chaparral,* (q. v.).

Dust Mulch

A shallow layer of loose surface soil.

Dust Whirl (Dust Devil)

A small, intense, vortical disturbance, usually only a few yards in diameter, in which large volumes of dust and debris are carried upward; occur usually in arid and semi-arid regions.

Dynamic Equilibrium

A system that is maintained in approximately the same condition because of the action of opposing processes or activities proceeding at about equal rates. cf. *Balance of nature.*

Dynamics, Community

See *Community dynamics.*

Dynamics, Population

See *Population dynamics.*

Dysgenic

Refers to any influence that is detrimental to the genetic properties of a population. cf. Eugenics.

Dystrophic

A type of lake or pond which contains brown water with much humic material in solution and with a small bottom fauna characterized by pronounced oxygen consumption. Cf. *Eutrophic.*

E

Ecad

A habitat form; an organism showing *Somatic* adaptations to a specific environment, not hereditable, cf. *Phenotype, Ecotype.*

Ecesis

See *Establishment, Spread, Invasion.*

Echard

A term occasionally used to denote the water in the soil below the permanent *Wilting percentage* (q. v.); not available for absorption by plants.

Echinoderm

A marine animal in the phylum Echinodermata such as starfish, sea-cucumbers, and sea-urchins.

Echolocation

The ability of certain animals, especially bats, to orient themselves by emitting high-frequency sounds and detecting their echoes; acoustic orientation.

Ecize

To undergo *Ecesis* (q. v.).

Ecocline

(1) A gradation or *Cline* (q. v.) in the adaptations of a species that is associated with an environmental gradient, cf. *Geocline.* (2) A gradation of ecosystems along an environmental gradient, comprising both the gradient of natural communities (*Coenocline*) and the *Complex gradient* of environmental conditions.

Ecological Amplitude

The range of one or more environmental conditions in which an organism or a process can function. cf. *Tolerance, Optimum, Pessimum.*

Ecological Bonitation

The estimate of the numerical abundance of an organism in a locality or a season. cf. *Bonitation, Biotic potential.*

Ecological Efficiency

The ratio between the energy available to one or more organisms or processes and the energy that is actually utilized.

Ecological Equilibrium

See *Balance of nature, Dynamic equilibrium.*

Ecological Equivalence

The situation or condition in which two or more species because of their similarity in *Ecological amplitude* can occupy the same ecological *Niche,* thus being able to replace each other.

Ecological Equivalent

An organism which participates in *Ecological equivalence* (q. v.). cf. *Vicariation.*

Ecological Factor

Any part or condition of the environment that influences the life of one or more organisms; often classified into A;

climatic, physiographic and edaphic, and biotic factors, or B; direct, indirect, and remote factors. cf. *Limiting factor.*

Ecological Indicator
See *Indicator.*

Ecological Longevity
The average length of life of individuals of a population under stated conditions. cf. *Life-span.*

Ecological Niche
See *Niche.*

Ecological Pyramid
See *Pyramid of numbers.*

Ecological Race
See *Ecotype.*

Ecological Sociology
See *Synecology.*

Ecological Structure
See *Structure.*

Ecological Succession
See *Succession.*

Ecological Valence
See *Ecological amplitude.*

Ecology
The study of the interrelationships of organisms to one another and to the environment. cf. *Autecology, Synecology, Bioecology, Sociology, Plant sociology.*

Ecophene
See *Ecad.*

Ecospecies
A *Taxon* of plants consisting of one or more *Ecotypes*

(q. v.) within a *Coenospecies* (q. v.), capable of reproduction, approximately equivalent to *Species* (q. v.).

Ecosystem

The *Community* (q. v.), including all the component organisms together with the abiotic environment, forming an interacting system, e. g., a marsh. cf. *Biogeocenose.*

Ecotone

A transition line or strip of vegetation between two communities which has characteristics of both kinds of neighboring vegetation as well as characteristics of its own. cf. *Edge-effect.*

Ecotype

The smallest *Taxon* (q. v.) or group of similar *Biotypes* (q. v.) within an *Ecospecies* (q. v.), each one adapted to a certain combination of environmental conditions. Differences between ecotypes may be morphological, or only physiological. cf. *Habitat form, Ecad.*

Ectoparasite

A *Parasite* living on the outside surface of another organism, e. g., a flea. cf. *Endoparasite.*

Ectophagous

Refers to an organism that feeds from the outside of the structure it is consuming such as a deer feeding on leaves of a plant. cf. *Entophagous.*

Ectotrophic

Refers to fungi that grow on the surface of roots. cf. *Endotrophic, Mycorrhiza.*

Edaphic

Refers to the soil, cf. *Edaphic factor.*

Edaphic Climax

See *Climax.*

Edaphic Factor

A condition or characteristic of the soil, physical, chemical, or biological that influences organisms. cf. *Biotic, Climatic, Ecological factor.*

Edaphology

The study of soils.

Edaphon

The aggregate of organisms in the soil except the roots or underground stems of plants. cf. *Plankton.*

Edge Effect

The influence of two communities upon their adjoining margins or fringes, affecting the composition and density of the populations in these bordering areas, e. g., a forest edge bordering a grassland. cf. *Ecotone.*

Effective Temperature Range

The range between the highest and the lowest temperature in which an organism can live. cf. *Ecological amplitude, Tolerance.*

Effluent

The outflow of water from subterranean storage. cf. *Influent.*

Elaioplast

A *Plastid* in which oil is formed and stored.

Electrolyte

Salts, acids, or bases that in a solution conduct an electric current, e. g., sodium chloride dissolved in water.

Element

(1) Organisms that are typical or characteristic of a certain region, but may occur outside of it, e. g., a group of prairie species occurring in the eastern part of the United States. (2) See *Ecological factor.*

Elementary Species
See *Ecotype.*

Elfin Forest
See *Krummholz.*

Eltonian Pyramid
See *Pyramid of numbers.*

Eluvial Layer
See *A horizon.*

Eluviation
The removal of material from a soil horizon by downward or lateral movement in solution and to a lesser degree in colloidal suspension. cf. *Illuviation.*

Emasculation
In plants the removal of anthers or flowers bearing stamens to prevent self-pollination. In animals the removal of sperm-producing organs.

Embryo Sac
The structure within the ovule of a flowering plant, in which *Fertilization* occurs.

Emigration
The *Migration* of an organism out of a locality, usually without the probability of returning. cf. *Immigration.*

Encinal
A grove or forest of evergreen oaks.

Enclosure
An area fenced to include certain kinds of animals. cf. *Exclosure.*

Encystment
The state of inactivity of an organism in which it is surrounded in a protective case; metabolism is reduced, resistance to unfavorable environmental conditions increased.

Endemic

A *Taxon* confined to a certain country or region and with a comparatively restricted distribution.

Endemism

The occurrence of endemics in an area.

Endobiophyta

The existence of a *Parasite* within the body of another organism.

Endocrine Gland

A gland in animals that produces hormones, e. g., the *Pituitary* gland.

Endogenous

Refers to a substance or process that originated within an organism or a cell.

Endoparasite

A *Parasite* living inside of another organism, e. g., a tapeworm.

Endophyte

A plant which grows within another plant such as a fungus *Endoparasite;* a plant which can penetrate a rock, e. g., lichen.

Endoskeleton

The supporting framework inside the body of animals such as *Vertebrates.* cf. *Exoskeleton.*

Endosperm

The nutritive tissue that surrounds the growing embryo, and which is present in the mature seed in many kinds of *Spermatophytes* such as the grasses.

Endotrophic

Refers to fungi that grow within roots. cf. *Ectotrophic, Mycorrhiza.*

Endozoochore

A *Propagule* such as a seed or a spore which is dispersed by being transported inside of an animal's body.

Energy Flow

The intake, conversion, and passage of energy through organisms or through an *Ecosystem* (q. v.).

Energy Transformers

Plants and animals which convert and pass on energy, originally secured from sunlight by plants, from one organism to another as in a food-chain. cf. *Energy flow.*

Entomophilous

Refers to plants that are pollinated by insects, e. g., orchids. cf. *Anemophilous.*

Entophagous

Animals that feed inside of dead leaves and roots. cf. *Ectophagous.*

Entropy

The degradation of energy, a measure of the degree of disorder of a system.

Environment

The sum total of all the external conditions which may influence organisms. cf. *Habitat, Site.*

Environment, Fitness of

The suitability of an environment or habitat for maintaining life.

Environmental Clock

The influence of environmental factors to initiate processes or activities of organisms, e. g., the initiation of flowering in the cocklebur by the *Photoperiodic* influence of short days and long nights.

Environmental Conditioning
See *Conditioning, environmental.*

Environmental Form
See *Ecad.*

Environment, Holocoenotic
The concept that the environmental factors act as a whole or aggregate in their effect upon one or more organisms.

Environmental Resistance
The restriction caused by environmental factors upon the increase in numbers of individuals in a population. cf. *Biotic potential, Reproductive potential.*

Enzyme
An organic catalyst, produced by living cells, each kind determining a specific chemical reaction, e. g., diastase which digests starch.

Eocene
The second geological epoch in the Cenozoic Era, Tertiary period, began about 58 million years ago and lasted for about 19 million years.

Eolian
See *Aeolian.*

Epeirogenesis
Great alterations in the level of the crust of the earth such as the elevation or lowering of the surface of continents.

Epharmone
An organism which has been subjected to *Epharmony* (q. v.). See *ecad.*

Epharmony

The acquirement by an organism of processes or morphological structures by which it is enabled to exist in an altered environment. cf. *Adaptation.*

Epharmose

The process of gradual adaptation of a species to a change of environment.

Ephemeral

Refers to short-lived existence.

Epibiotic

An endemic surviving from a former flora, a *Relic.*

Epicole (Epibiont)

An organism living attached to another organism without benefit or harm to the latter, e. g., barnacles attached to corals, algae on the bark of trees. cf. *Commensalism, Epiphyte.*

Epidemic

The widespread occurrence in greater numbers than usual of a species that is usually parasitic or predatory.

Epidermis

The outermost layer of cells of animals and plants; cork cells replace the epidermis in stems and roots of older woody plants.

Epigeal

Refers to an organism that lives close to the ground, e. g., some insects, cotyledons on seedlings such as the navy bean. cf. *Hypogeal.*

Epilia

A community of *Epiphytes* (q. v.).

Epilimnion

The upper layer of lakes, subject to disturbance by winds, lying above the *Thermocline,* (q. v.). cf. *Hypolimnion.*

Epinasty

The downward curvature of a plant organ such as a leaf, caused by the greater growth on the upper surface than on the lower.

Epiorganism

A natural group or entity consisting of individual organisms, e. g., a society of termites, a *stand* or *community-type*. syn. supraorganism.

Epiphyll

An *Epiphyte* (q. v.) growing on a leaf, e. g., certain lichens.

Epiphyte

A plant growing upon or attached to another plant, or often on some non-living support, deriving no sustenance from the supporting structure, e. g., Spanish moss on a live oak.

Epiphyton

An assemblage of organisms scattered on submerged surfaces which later may become mechanically associated. cf. *Periphyton, Lasion.*

Epiphytotic

An *Epidemic* disease in plants, e. g., wheat rust.

Epithalassa

The upper layer, above the *Thermocline,* where *Thermal* stratification occurs in the ocean.

Epizoan

A non-parasitic animal living attached to another. cf. *Commensalism.*

Epizoochore

A *Propagule* such as a spore or a seed that is carried on the body of an animal.

Epizootic

An Epidemic disease in animals. cf. *Epiphytotic.*

Equilibrium, Community

The condition in which a community is maintained with only minor fluctuations in its composition within a certain period of time.

Equilibrium, Ecologic

See *Balance of nature, Dynamic equilibrium.*

Equivalence, Ecologic

See *Ecological equivalence.*

Eremean (Eremic)

Refers to desert vegetation.

Eremophilous

Refers to organisms living in deserts.

Eremophyte

A plant growing in a desert.

Erg

The part of the Sahara desert covered with sand dunes.

Ergot

A disease of cereals, especially rye, and other grasses, caused by the fungus *Claviceps purpurea;* in which dark-colored structures replace the grain.

Ericaceous

Refers to the heath family of plants, Ericaceae.

Ericad

A member of the heath family, Ericaceae, e. g., blueberry.

Ericilignosa

A type of vegetation in which ericaceous plants are dominant or very abundant.

Erodible

A substance, especially soil, that is susceptible to erosion. cf. *Erosive.*

Erosion

The detachment and movement of particles of the land surface by wind, water, ice, or earth movements such as landslides and creep. cf. *Accelerated erosion, Normal erosion, Sheet erosion.*

Erosion Class

One of several categories in a classification indicating the degree of erosion.

Erosion Pavement

A layer of stones or gravel on the surface of the ground after fine particles have been removed by erosion.

Erosive

Refers to the tendency of an agent such as water or wind to cause erosion. Erosive is preferred when referring to the agent, *Erodible* when referring to the substance that is eroded.

Escape

A plant found wild but which originated from a cultivated ancestor.

Escape Covert

Vegetation which is intended or used for protection by animals from attack by enemies.

Escape Mechanism

A structure, behaviour, or process that enables an organism to survive unfavorable conditions, e. g., shedding of leaves, burrowing, development of cysts.

Escarpment

A long, inland cliff or steep slope, usually high, formed by erosion or possibly by faulting. syn. scarp.

Esker

A long, narrow ridge of gravel and sand deposited by a stream flowing under or within a glacier. cf. *Kame.*

Esophagous (Oesophagous)

The portion of the alimentary tract between the pharynx and the stomach.

Espinal

A thorny woodland.

Essential Element

A chemical element required by green plants for normal growth, such as the primary essential elements: hydrogen, oxygen, nitrogen, phosphorus, and potassium; secondary essential elements: sulfur, calcium, and magnesium; and the trace or minor elements: iron, boron, manganese, copper, zinc, and molybdenum. The last six, and traces of other elements, are required in only minute quantities.

Establishment

The successful growth of an organism in a new location. syn. *Ecesis.* cf. *Invasion, Spread, Dispersal.*

Estival

Refers to the summer season, cf. *Aspection.*

Estivation

The condition in which an organism may pass the summer and in which its normal activities are greatly curtailed or temporarily suspended. cf. *Dormancy, Hibernation.*

Estuary

An arm of the sea at the mouth of a river, in which the current of the river meets the tide.

Ethiopian Region

The faunal region in the realm *Megagea* (Arctogea); it includes all of Africa except the northern corner and part of southern Arabia.

Ethnobotany

The study of the use of plants by any race or group of people. cf. *Paleobotany, Paleontology.*

Ethnology

The study of the distribution and characteristics of the divisions of mankind.

Ethology

The comparative study of the *Behaviour* of animals.

Etiolation

The development of a plant grown without light, resulting in the loss of chlorophyll, a weak elongated stem, and abnormal leaves.

Etiology

The study of the causes of diseases.

Eucoen

See *Index species.*

Eugenics

The study of the improvement of the genetic constitution of a population or species, especially the human race.

Eulittoral

The shoreward part of the *Benthic zone* in lakes and oceans; the landward part of the littoral zone, including all of the intertidal region. cf. *Littoral.*

Euphotic Zone

The uppermost portion of a body of water which receives sufficient light for *Photosynthesis.* cf. *Aphotic, Disphotic, Photic zones.*

Euplankton

Animal components of the *Plankton* (q. v.).

Euploid

Refers to the presence of a regular number of chromosomes, a multiple of the *Haploid* number. cf. *Aneuploid.*

Euroky

The capacity of an organism to live under a wide range of environmental conditions. syn. *Eurytopic.* cf. *Ecological amplitude, steno-.*

Eurytopic

See *Euroky.*

Euthenics

The study dealing with the improvement of living conditions in order to secure better human beings. cf. *Eugenics.*

Eutrophic

Refers to bodies of water, accumulations of peat, etc., which are rich in mineral nutrients and organic materials, therefore productive. Oxygen may be deficient seasonally in lakes or ponds. cf. *Oligotrophic, Dystrophic.*

Evaporative Power of the Air

The environmental factor complex including factors such as temperature, relative humidity, and wind that influence the evaporation of water from organisms and from other bodies containing water.

Evaporimeter

An instrument to measure the *Evaporative power of the air* such as an open pan of water or an *Atmometer* (q. v.).

Evapotranspiration

The sum total of water lost from the land by evaporation and plant *Transpiration.*

Evapotranspirometer

A soil-filled container covered with living vegetation and supplied with water to measure transpiration and evaporation.

Even-aged

Refers to a stand of trees in which only small differences in age occur between the individuals; in young stands differences should not exceed 10 to 20 years, in mature stands not more than 30 to 40 years. cf. *Age-class.*

Evolution

The process of *Natural* or *Artificial selection* acting upon genetic diversity in organisms.

Evolution, Convergent

The development during the course of evolution of similar structures or habits in organisms that are not closely related taxonomically. cf. *Taxonomy.*

Exchange Capacity

See *Cation-exchange capacity.*

Exclosure

An area fenced to exclude certain kinds of organisms. cf. *Enclosure.*

Exclusive Species

A species belonging to the highest *Fidelity* class (Braun-Blanquet), one which occurs exclusively or almost so in a specific kind of vegetation unit.

Exclusiveness

The degree in which a particular species is restricted to a particular kind of community to the exclusion of other communities. cf. *Exclusive species, Fidelity.*

Excretion

The elimination of substances from the bodies of organisms.

Excurrent

A plant with one main stem, the lower branches longer than the upper, e. g., a spruce tree.

Exempt Stock

Livestock such as work or saddle horses and milk cows which are permitted to graze on National Forests or the Public Domain free of charge.

Exfoliation

The peeling off of material in thin layers from the surface of rocks or the flaking off of scales or other parts of organisms.

Exoskeleton

A hard, supportive structure in the outer part of the body of many kinds of animals, e. g., shell of a crab.

Exothermic

Refers to the release of heat.

Exotic

Refers to any organism that is not native in the area where it occurs; introduced. cf. *Endemic, Indigenous.*

Experimental Design

The plan of an experiment, especially to insure that the data to be secured will be suitable for statistical analysis.

Exploitation

The ability of an organism to find, occupy, and retain unused vital resources. cf. *Competition, Interference.*

Exsiccata

Dried, preserved specimens of plants.

Exsiccation

The process of promoting the development of desert conditions through human activity or climatic change. cf. *Siccation.*

Extracellular

Refers to something outside of a cell, but may be within a multicellular organism. cf. *Intracellular.*

Extraclinal

Refers to an organism or a population that is not part of a *Cline* (q. v.).

Extraneous

Refers to the presence of an organism near the border of its range or nearer the margins than the *Center of dispersal.*

Extrinsic Cycle

See *Cycle.*

Eyrie

The nest or brood of a bird of prey such as an eagle or raven.

F

Faciation

A part of a *Climax Association* (sensu Clements) which lacks some of the dominants of the association because of slight differences in environmental conditions. cf. *Lociation.*

Facies

(1) See *Faciation.* (2) The general appearance or aspect of a plant, population, or community. cf. *Aspection.* (3) A modification of the *Biotope,* differing recurrently from the typical conditions in minor ways. (4) A variation in a community such as a dogwood or mountain laurel shrub facies in an oak forest.

Facilitation, Social

The effect of the presence of one organism upon the behaviour of another.

Factor, Ecological

Any part of the environment that influences the life of an organism. cf. *Biotic, Limiting, Density-dependent factors.*

Facultative

Refers to the capability of an organism to live under

various conditions such as parasitism and saprophytism. cf. *Obligate parasite.*

Fairy Ring

A circle of mushrooms arising from underground mycelial growth, usually accompanied by a luxuriant ring of vegetation, fairly common in grasslands.

Fall Line

A line connecting the points where rivers leave the uplands for the lowlands, marked by an increased slope and waterfalls, e.g., the Atlantic coastal plain adjoining the Appalachian Mountains.

Fallout

The dropping to the earth from the air of solid material, particularly the radioactive dust from atomic explosions.

Fall Overturn

See *Spring overturn.*

Fallow

Refers to cropland left idle except for tillage in order to destroy weeds and accumulate water and nutrients for use of a crop to be planted later. cf. *Summer fallow.*

False Annual Ring

More than one of the *Growth rings* that may be produced in woody plants in a single season.

Family

In the classification of organisms a group of one or more related genera, e.g., the rose family in which the roses, strawberries, etc., are classified. In ecology it has been used for a community comprising individuals of a single species.

Fasciation

The condition in plants in which stems and branches have grown together to form a malformed and flattened structure.

Fascicle

A dense cluster or bundle such as three to five leaves in groups on pine trees.

Fauna

A collective term to include all the kinds of animals in an area or in a geological period.

Faunal Region

An area containing characteristic kinds of animals, e.g., the *Oriental Region,* (q. v.).

Fecundity

Capability of an organism to produce reproductive units such as eggs, sperms, or asexual structures. cf. *Fertility.*

Feed

Harvested forage including hay or fodder, or grain, grain products, and other foodstuffs that are processed for feeding livestock.

Feedback

The return of a substance to a former condition or location, such as the absorption of calcium by plants and the return to the soil when the leaves fall to the ground and decay.

Fell, Fell-field

A tract of bare, elevated country which is in more or less uncultivated condition.

Felling

In selective felling only certain trees are cut down in a forest, in clear felling all are taken.

Fen

A tract of low, marshy ground containing peat, relatively rich in mineral salts, alkaline in reaction, situated in the upper parts of old estuaries and around fresh-water lakes, vegetationally distinct from *Moors* (q. v.).

Feral

Refers to an organism that has escaped from cultivation or domestication and become wild again.

Fermentation

The chemical alteration of organic substances by organisms, especially yeast and bacteria, e.g., alcoholic fermentation in which sugar is changed to alcohol and carbon dioxide by the enzyme zymase, produced by yeast.

Fertile

Refers to the capability of an organism or organ such as a stamen, pistil, or ovary to carry on reproductive functions.

Fertility

(1) The capability of an organism to produce living offspring. cf. *Fecundity*. (2) The quality of a soil which enables it to provide substances in adequate amounts properly balanced and available for the growth of specified plants when other environmental factors are favorable.

Fertilization

The union of egg and sperm, or of two *Gametes* (q. v.).

Fertilizer

Any material added to soil to supply one or more plant nutrients, usually not including lime or gypsum.

Fibrous Root System

The aggregate of the numerous, similar roots arising from the base of the stem such as occurs in corn and the grasses. cf. *Tap root system.*

Fidelity

The degree of regularity or "faithfulness" that a species occurs in certain plant communities, expressed in a 5-part scale: (5) *Exclusive,* (4) *Selective,* (3) *Preferential,* (2) *Companion,* indifferent, (1) *Accidental,* strangers (q. v.).

Field Border Plantings

Vegetation established on the borders of fields to conserve soil and to provide cover and food for wildlife, e.g., a strip of multiflora rose.

Field Capacity

The moisture content of a soil, expressed as the percentage of oven-dry weight (100-110° C.) after the *Gravitational water* has drained away; the field moisture content two or three days after a soaking rain. cf. *Capillary water, Water-holding capacity.*

Field Crops

Crops such as grain, hay, root, and fiber in contrast to vegetable (truck) and fruit crops.

Field Layer

The stratum of vegetation formed by grasses, forbs, and dwarf shrubs.

Field Strip Cropping

The kind of *Strip cropping* in which crops are grown in parallel strips across a slope but which do not follow contour lines, and strips of grass or other close-growing plants are grown alternately with the strips of cultivated crops.

Field Test

An experiment conducted under ordinary field conditions, usually less subject to control than a precise experiment.

Filament

The stalk of an *Anther* or of a down-feather.

Filter Bridge

A land connection, temporary in duration and restricted in extent, limiting the kinds of organisms which can migrate over it, e.g., the Bering Strait in the *Pleistocene.* cf. *Corridor, Sweepstakes bridge.*

Filter Strip

A strip of permanent vegetation sufficiently wide and dense above a farm pond, diversion terrace, etc., so it will retard run-off and cause deposition, thus preventing silting in the water or structure below.

Fine-textured Soil

A soil that consists mainly of silt and clay.

Fiord (Fjord)

A narrow, often long and deep inlet of the sea as in the coasts of Norway and Alaska, very likely formed by glaciers.

Firebreak

A strip of land from which inflammable materials have been removed in order to check or stop a creeping or running fire.

Fire Control Line

The line along which efforts are made to stop the advance of a fire or from which to start to backfire.

Fire Hazard

The risk of probabilities that a fire may start because of the inflammability of materials under the prevailing climatic conditions.

Firn

See *Névé*.

First Bottom

The *Flood plain* that is most immediate to a stream, or at the lowest elevation above a stream, consequently some are frequently flooded.

Firth

A long, narrow arm of the sea or lower portion of an estuary, especially in Scotland.

Fishway

A sloping structure over which water flows, to enable fish to ascend a stream around a dam or other obstruction; a fish ladder.

Fission

(1) Reproduction of a unicellular organism by simple division into two parts. (2) Refers to atomic bombs in which elements such as uranium and plutonium are split into products as, for example, Sr^{90}, I^{131}, and Cs^{137}, which are formed during the explosion.

Fitness

The degree of *Adaptedness* (q. v.) to the environment that an organism possesses.

Fitness of the Environment

The suitability of environmental conditions such as the nature of the water, gaseous composition of the atmosphere, and temperatures for the maintenance of life or for the activities of a certain organism.

Fixation (in Soil)

The conversion of a soluble substance such as phosphorus from a soluble or exchangeable form to a relatively insoluble form.

Fjaeld (Fjeld)

A more or less barren upland area (Scandinavia). See *Fell.*

Flachmoor

A moor with a flat or even slightly concave surface and soil poor in salts and acid in reaction.

Flatworm

A member of the phylum Platyhelminthes, e.g., flukes, tapeworms.

F Layer (Soil)

Sometimes used to designate the partially decomposed litter which can still be recognized as to origin and age. cf. *A horizon, H layer, L layer.*

Flocculation

The aggregation of suspended colloidal material or very fine particles into larger masses or floccules.

Flood Plain

The nearly level land forming the bottom of a valley in which a stream is present and usually subject to flooding. cf. *First flood plain.*

Floodway

A channel usually bordered by levees for the purpose of carrying flood water.

Flora

The sum total of the kinds of plants in an area at one time. cf. *Vegetation, Fauna.*

Floral Region

See *Floristic region.*

Floret

The small flower of grasses or *Composites.*

Florigen

A hormone evidently made in the leaves which after translocation to apical *Meristem* initiates the formation of flowers.

Floristic Area

An area in which a certain degree of homogeneity exists because of similarities in the areas occupied by various species.

Floristic Composition

The kinds of plant species, in the aggregate, that occur in a community or in an area.

Floristic Element

Species that are characteristic of a certain territory but occur also in a different one, e.g., an arctic species growing in the high Rocky Mountains is an arctic element in the flora of the Rockies.

Floristic Region

A portion of the earth's surface supporting a characteristic flora which developed largely within this portion, e.g., the Atlantic North American Region of the Boreal Kingdom.

Floristic Territory

A geographic area characterized by the common occurrence of a number of species which are more or less confined to it, but other species may also be present.

Flower

The organ of the class Angiospermae, flowering plants, consisting of one or more pistils (carpels) or stamens, or both, and usually a calyx (composed of sepals) and a corolla (composed of petals).

Flower Induction

The stimulation, presumably by *Florigen* (q. v.) which initiates the production of flowers.

Fluctuation

A relatively irregular departure from more normal or average conditions. cf. *Community dynamics.*

Fluke

A parasitic flatworm in the class Trematoda, phylum Platyhelminthes, e.g., the liverfluke in sheep.

Flume

An open conduit of wood or other material for carrying

water or other liquids across a creek, ravine, or other depression which lies across the course of a canal or ditch location.

Flyway

A migration route of birds, e.g., the Mississippi River flyway.

Fodder

The dried, cured plants of crops such as corn and sorghum, including all parts above ground including the grain. cf. *Stover, Hay.*

Foehn

A dry wind which is warm for the season. It blows down leeward slopes of mountains, especially in the Alps. cf. *Chinook.*

Fog

The condensation of water vapor on particles of dust or smoke particles.

Foliaceous

Refers to structures that are leafy or leaf-like, thin.

Foliage Cover

See *Cover.*

Foliar Diagnosis

Evaluation of the status of plant nutrients in a plant, or the plant-nutrient requirements of a soil, by the analysis of leaves.

Follicle

A dry, many-seeded fruit developed from a simple ovary which when ripe splits along a single line, e.g., the larkspur fruit.

Food-chain

A figure of speech for the dependence for food of organisms upon others in a series, beginning with plants or scavenging organisms and ending with the largest carnivores.

Food-cycle

All the interconnecting *Food-chains* in a community. syn. food-web.

Food-niche

The particular location of an organism in a *Food-cycle*.

Food-pyramid

A graphic representation of the food chain which indicates the large numbers of producer organisms at the base and the progressively decreasing numbers of herbivores and carnivores above.

Food-web

See *Food-cycle*.

Forage

Unharvested plant material available as food for domestic animals, may be cut for hay or grazed; after cutting it is called *Feed. cf. Fodder.*

Forage-acre

A theoretical concept, the quantity of forage on an acre of land which is completely covered with herbage and completely utilized under proper management.

Forage-acre Requirement

The number of forage acres needed for the maintenance of a mature grazing animal for a certain period of time.

Forage Fish

Small kinds of fish which reproduce prolifically and are used as prey by predatory fishes.

Forage Ratio

This is the ratio of the percentage of a certain prey

organism present in the food of a predator divided by the percentage of this prey that is present in the habitat.

Forage Volume

(1) The parts of a plant within reach of animals for grazing. (2) A measure of the yield of *Forage,* the total amount of forage produced on a range area during a year.

Forb

An herbaceous plant that is not a grass nor grasslike such as a sedge, e.g., sunflower, geranium.

Foredune

A low dune, often occupied by a sand-binding grass, bordering the sandy shore of a sea or lake.

Foreshore

The portion of the shore occupied daily by tides.

Forest

A stand of trees growing close together with associated plants of various kinds.

Forest Cover

The living plants and dead organic material occupying the surface of a forest, often restricted to the woody plants covering the ground. cf. *Basal cover.*

Forest Edge

The border, or *Ecotone,* of a forest with another kind of vegetation such as grassland.

Forest Floor

The deposits of plant material such as leaves and branches on the ground in a forest.

Forest Influences

All the effects or *Reactions* of a forest upon the habitat or the environmental conditions, e.g., aiding in maintenance of uniform stream flow, shading the ground.

Forest Type

A forest stand that is essentially similar throughout its extent in composition under generally similar environmental conditions. It includes temporary, permanent, climax, and cover types.

Form

A botanical taxonomic category based upon more trivial characteristics and with a less distinct geographical range than those of the *Subspecies* or *Variety.*

Formation

One of the largest subdivisions of the vegetation of the earth, usually of great geographical extent, composed of communities that are similar in physiognomy and broad environmental relations, e.g., the deciduous forest of eastern North America. cf. *Climax, Biome.*

Formenkreis

A series of related forms distinguished geographically and which originated entirely or primarily by geographic isolation.

Fossorial

Refers to animals that burrow in the soil, e.g., the mole. cf. *Ambulatorial.*

Fragment (of a Community)

A stand so small that it lacks the species composition and other characteristics of the *Community.*

Fragmented Structure (Soil)

A soil composed largely of particles that have well defined faces and edges. cf. *Granular structure.*

Free-martin

A sexually imperfect female calf, sterile usually, born as a twin of a male animal.

Frequence

See *Frequency.*

Frequency

The degree of uniformity with which individuals of a species are distributed in an area, and more specifically in a *Stand.* cf. *Constancy.*

Frequency Class

One of the small groups into which the *Frequency indices* of the various species in a stand may be classified.

Frequency Index

The quantitative expression in percentage of *Frequency,* e.g., a species occurring in 15 of 20 sample areas in a stand has a frequency index of 75 per cent.

Frequency, Law of

The generalization which states that when *Frequency indices* of species in a stand are classified into five main classes a double peak occurs in homogeneous vegetation, i.e. $A > B > C \quad D < E$.

Frigid Zone

The portions of the earth north of the Arctic Circle and south of the Antarctic Circle. Cf. *Temperate zone. Tropics.*

Frigorideserta

Tundra or cold arctic and alpine, open communities.

Fringing Forest

A strip or zone of forest along a stream bed. cf. *Gallery forest.*

Front

The border of cold and warm air masses at the earth's surface.

Frost

(1) The act or state of freezing, or injury to organisms

because of low temperatures, especially near the beginning or the end of the growing season. (2) Particles of frozen water or dew (hoarfrost) appearing on the earth's surface at 32°F. or lower.

Frostless Season

The period between the last frost in the spring and the first one in the autumn.

Frost Resistance

The capability of plants to survive the formation of ice crystals in their tissues.

Fruit

The ripe ovary or group of ovaries with any other parts that may be regularly associated with it, e.g., a grain of corn, a gooseberry fruit, an apple pome.

Frutescent

Resembling a shrub somewhat. See *Fruticose.*

Fruticeta

Vegetation types consisting of scrub forests.

Fruticose

Shrubby, cf. *Suffruticose.*

Fruticose Lichen

A *Lichen* with a *Thallus* ten cm. or more tall, e.g., *Cladonia rangiferina,* the reindeer "moss."

Fucoid

Resembling a seaweed, especially Fucus.

Fugacious

Refers to a structure that falls or separates early from a plant, e.g., sepals or petals of some flowers.

Fully Stocked

Refers to a stand which contains as many trees or as much

material of the species and of the age as the *Site* can support. Cf. *Overstocked, Understocked.*

Fumarole

A hole in the earth from which gases such as CO_2 and steam escape under pressure, as seen in the Valley of Ten Thousand Smokes, Alaska.

Fungicide

A substance that is toxic to fungi, e.g. Paris green.

Fungivorous

Refers to organisms which consume fungi, e.g., some *Collembolons* and mites.

Fungoid

Resembling a fungus.

Fungus

One of the true fungi, belonging to the phylum Eumycophyta; plants lacking *Chlorophyll* such as the molds, yeasts, mildews, rusts, and mushrooms. They may be either *Parasitic* or *Saprophytic.*

Funiculus

The stalk of the *Ovary* in plants.

Furrow Dam

A small earth dam for the purpose of holding water within a furrow. cf. *Lister.*

Fynbos

Sclerophyllous vegetation on plateaus and mountains in South Africa, ecologically equivalent or similar to *Macchia* of the Mediterranean region and the *Chaparral* of California.

G

Gallery Forest (Galleria)

Woods or a narrow zone of forest along a stream in grassland, savanna, or other open vegetation.

Game Management

The practice of producing sustained annual crops of wild game on land. cf. *Range management.*

Game Refuge

An *Enclosure* for the purpose of controlling or prohibiting the hunting, fishing, or otherwise destroying game animals, birds, fish, and other animals.

Gamete

A sex cell, a sperm or egg; in some of the simplest organisms the gametes are not differentiated into egg and sperm.

Gametocide

An *Herbicide* that prevents development of or destroys *Gametes.*

Gametophyte

The plant or generation in organisms that produces gametes, contains the *Haploid* number of *Chromosomes.*

Gamma Radiation
One kind of ionizing radiation, electromagnetic, readily penetrates biological materials.

Gamopetalous
Refers to flowers that have more or less united *Petals.*

Gamosepalous
Refers to flowers that have more or less united *Sepals.*

Garique (Garrique)
Open vegetation consisting of dwarf, evergreen *Scrub* on poor land in the Mediterranean region. cf. *Maquis.*

Gastropod
An animal in the class Gastropoda, phylum Mollusca, e.g., slug, snail.

Gause's Principle
The generalization that states that two species do not occupy exactly the same *Niche.* See *Competitive exclusion principle, Grinnell's axiom.*

Geiger Counter
An instrument for detecting *Ionizing radiation.*

Geitonogamy
The *Pollination* of a flower by pollen from another flower on the same plant. cf. *Autogamy, Cleistogamous, Xenogamy.*

Gemma
A structure consisting of a few cells which becomes separated from the parent and grows into a new plant, found in certain liverworts.

Gene
A localized unit of genetic material with a specific function in transmitting characters from one generation to the following one.

Genecology

The study of population genetics in relation to th habitat conditions; the study of species and other taxa by th combined methods and concepts of ecology and genetics.

Gene Flow

The duplication and dispersal of genes in a population

Gene Pool

In a narrow sense, the genic material of a local inter breeding population at the present time. In a broad sense the total genic resources or materials of a species throughou its geographical range.

Generic

Refers to *Genus.*

Generic Coefficient

The ratio of the number of genera to the number o species in an area.

Genetic Drift

Accidental fluctuations in the proportions of a particula *Allele* so that exact Mendelian ratios do not occur, whic may result in the fixation or loss of certain genes in smal populations without reference to selective value. cf. *Natura selection.*

Genetics

The branch of biology dealing with *Heredity* in all i phases.

Genome (Genom)

The set of different *Chromosomes,* as found in a *Gamet* or *Haploid* nucleus; the *Diplod* nucleus contains tw genomes, *Polyploids* more than two.

Genotype

The entire genetic constitution, or the sum total of gene of an organism, in contrast to the *Phenotype.*

Genus

A group of related species, or occasionally only one species, used in the classification of organisms, e.g., the white and Scotch pines belonging to the genus *Pinus*.

Geobiont

An organism that spends all its life in the soil, e.g., certain fungi, protozoa, and nematodes.

Geochronology

The study of biological and meteorological events in relation to time.

Geocline

A gradation or *Cline* (q. v.) based on topographic or spatial separation. cf. *Ecocline.*

Geocole

An organism that spends a portion of its life in the soil, e.g., certain mammals, reptiles, and other animals, cf. *Geobiont, Geoxene.*

Geoecotype

See *Geotype.*

Geographic Race

A *Race* restricted to a certain geographic area. cf. *Ecotype.*

Geological Erosion

See *Normal erosion.*

Geomorphology

The branch of physical geography that deals with the form and arrangement of the earth's crust.

Geophile

See *Geocole.*

Geophyte

One of Raunkiaer's *Life-form classes* of plants in which the buds or other *Perennating parts* surviving unfavorable seasons are buried in the surface soil, e.g., plants with bulbs, tubers, or rhizomes.

Geosere

A series of *Climax Formations* throughout geological time in an area.

Geotaxis

A *Taxis* (q. v.) in response to gravity.

Geotome

An instrument for securing soil samples.

Geotropism

A *Tropism* (q.v.) in response to gravity, e.g., the main roots of plants growing downward, the main stems upward.

Geotype

A genotypic population occurring in a habitat which is partly isolated by topographic barriers; includes most geographical *Races* and *Subspecies*. Syn. geoecotype. cf. *Genotype*.

Geoxene

An organism that occurs accidentally in the soil. cf. *Geocole*.

Germination

The process of growth renewal of a seed or spore; a seed is considered to have completed germination in some instances when the *Hypocotyl* projects outside of the seed coat, in other instances when the seedling appears above ground.

Germ Plasm

The protoplasm which transmits the hereditary characters or *Genes* (q. v.).

Gestation Period

The period of time that the embryo and fetus are in the uterus of an animal.

G Horizon
A soil layer developed wholly or partly in *Gley* (q. v.) soil, characterized by the presence of ferrous iron and usually by neutral gray colors.

Gigantism (Giantism)
A plant showing excessive vegetative growth.

Gilgai
The type of microrelief characterized by a succession of micro-basins and micro-knolls or many small ridges in nearly level areas on clay soil having high coefficients of expansion and contraction with changes in moisture content; "pits-and-mounds."

Girdling
The removal of a ring of bark or tissues from a stem, causing the death of the plant.

Glacial Drift
See *Drift*.

Glaciation
The covering of an area by a glacier or by an ice-sheet, or the geological action of the glacial ice upon the land.

Gleization
The soil-forming process in which *Gley* soil is formed.

Gley (Glei) Soil
A soil formed under the influence of water-logging and lack of oxygen; usually neutral gray in color and containing reddish brown deposits of ferrous hydroxide. cf. *G horizon*.

Gloger's Rule
The generalization which states that animals in warm climates tend to be darker in color than those in arid or cool regions.

Glycophyte

A plant growing in soil that is low in salt content, in contrast to *Halophyte,* e.g., American elm.

Graded Terrace

A terrace having a continuous slope along its length. cf. *Level terrace.*

Grain

An *Indehiscent* fruit in which the coat of the single seed is united with the ovary wall, e.g., wheat.

Gramineous

Refers to the grass family, Gramineae.

Graminivorous

Refers to an animal that eats grass.

Graminoid

Refers to an herb with long, narrow leaves.

Granular Structure (Soil)

A soil made up chiefly of particles or aggregates that have rather indistinct faces and edges. cf. *Fragmented structure.*

Grass

A plant in the family Gramineae with characteristically reduced flowers of florets, grain type of fruit, and with narrow, usually elongated leaves which are attached in two ranks to the jointed stem or culm.

Grassed Waterway

A natural or artificially made course for the flow of water, usually shallow, on which erosion-resistant grasses are grown, to permit water to run off fields thus reducing erosion where the crops are growing.

Grassland

Vegetation consisting chiefly of grasses or grasslike plants. cf. *Steppe, Prairie, Pampas, Meadow, Veld, Savanna.*

Grasslike Plant

A plant which resembles a true grass, e.g., sedges, rushes.

Gravitational Water

Water in large pores in the soil which drains away under the force of gravity when underdrainage is free.

Gray-Brown Podzolic Soil

A major soil group having a thin organic and thin organic-mineral layers over a grayish brown leached layer which rests upon a brown *B horizon* richer in clay than the horizon above; formed under deciduous forests in a moist temperate climate.

Grazing

The feeding by livestock and game animals on live or standing plants other than *Browse*.

Grazing Capacity

The maximum number of animals or animal units per acre, or acres per animal, that a grazing area can support without deterioration. cf. *Carrying capacity*.

Grazing District

An administrative unit of the Federal rangeland established by the Secretary of the Interior under provisions of the Taylor Grazing Act of 1934, as amended; or an administrative unit of a state, private, or other range lands, established under state laws.

Grazing Land

Land used regularly for grazing; not necessarily restricted to land suitable only for grazing but cropland and pasture used in connection with a system of farm crop rotation are usually not included. cf. *Range*.

Grazing Permit

An authorization for the grazing on public or other lands

under specified conditions, issued to a livestock operator by the agency in charge of the lands.

Grazing Preference

The criteria used in the administration of public grazing lands for the issuance of grazing permits and licenses.

Grazing Unit

A division of the *Range* for the facilitation of administration or the handling of livestock.

Green Manure Crop

A crop grown for the purpose of turning under while it is still green, or shortly after maturity, in order to improve the soil.

Gregariousness

The tendency of organisms to congregate or form groups, e.g., reindeer, cattails. See *Sociability*.

Grinnell's Axiom

The generalization stating that no two species in the same general territory can occupy for a long time the same ecologic *Niche*. cf. *Gause's principle*.

Ground Cover

See *Cover*.

Ground Water

Water standing in or moving through the soil and underlying strata, *Gravitational water*, the source of water in springs and wells. cf. *Runoff*.

Group Control

The control or influence of the behaviour of a group of animals, by means of the cue behaviour or a signal of specific animals.

Growing Point

One of the sections of a plant body in which growth is localized, especially the tips of stems and roots.

Growing Stock

The total number or total volume of all the trees in an area.

Growth Form

The characteristic shape or appearance of an organism as a result of its development in response to the impinging environmental conditions within its genetic constitution. cf. *Habitat form, Phenotype, Genotype.*

Growth Layer

A layer of *Xylem* and *Phloem* produced in woody stems usually during each growing season. cf. *Annual ring.*

Growth Ring

A *Growth layer* seen in the cross section of a woody stem.

Growth Substance

Any chemical produced by a plant, or synthetically, that regulates plant growth. cf. *Hormone.*

Gully

A channel or small valley formed by running water which usually flows only during and immediately after heavy rains or the melting of snow; it may be branched or linear and fairly long, narrow, and uniform in width; smaller than a ravine, deeper than a rill.

Gully Erosion

Removal of stones, gravel, and finer material by running water with the formation of channels that cannot be smoothed out completely by ordinary cultivation.

Guttation

The exudation of water in liquid form from plants through *Hydathodes,* (q. v.). cf. *Transpiration.*

Gymnosperm

A plant in the class Gymnospermae of the seed plants, *Spermatophytes,* in which the seeds are not enclosed within an ovary, e.g., pine, spruce. cf. *Angiosperm.*

Gynandromorph

An organism containing both male and female characteristics, e.g., certain insects. cf. *Hermaphrodite.*

Gynandrous

Refers to plants in which the stamens are fused with the pistil, e.g., certain orchids.

Gynoecium (Gynaecium)

The *Carpels* or *Pistils* of a flower considered collectively.

Gypsophilous

Refers to plants growing characteristically on soils rich in gypsum.

Gyttja

The jelly-like ooze on lake bottoms.

H

Habit

The general appearance of a plant such as tall and erect or decumbent and trailing. cf. *Life-form.*

Habitat

The sum total of environmental conditions of a specific place that is occupied by an organism, by a population, or a community. cf. *Environment, Niche, Site, Microhabitat, Standort, Station.*

Habitat Form

The *Growth form* or appearance of organism which is characteristic of a certain *Habitat.* cf. *Epharmony, Life-form, Ecotype.*

Habituation

The relatively waning response of an organism resulting from repeated stimulation which is not followed by any reinforcement.

Hadal Zone

The very deep part of the ocean, below 6000 meters. cf. *Abyssal.*

Halarch

Refers to saline conditions present in the soil at the beginning of a *Halosere* (q. v.).

Half-shrub

A perennial plant that is partly woody, usually at the base, and partly herbaceous, e.g., *Artemisia frigida.*

Halic

Refers to saline soil or to plants growing in such soil.

Halicole

A plant growing in soil with a high content of salts.

Halobiont

Any organism that grows in a *Saline habitat.*

Halomorphic

Refers to a soil found in poorly drained depressions in arid and semi-arid regions, cf. *Hydromorphic, Solonchalk, Solonetz.*

Halophyte

A plant growing in *Saline* soil.

Halosere

The series of stages in *Succession* originating on a *Saline* area. cf. *Sere, Hydrosere, Xerosere, Halarch.*

Hammada (Hamada)

Rocky uplands in deserts, bare of fine soil or sand because of wind action, used especially in the Sahara.

Hammock

A mesic area occupied by a community of hardwoods (Florida) or an island in a swamp (Okefinokee swamp).

Haplodiploidy

The condition occurring in some *Arthropods* in which males develop from unfertilized eggs and females from fertilized eggs.

Haploid

Refers to an organism or part of one in which the nuclei contain a single set of *Chromosomes,* i.e., one *Genome* (q. v.), e.g., *Spores, Gametes, Gametophytes,* certain male animals such as bees. cf. *Diploid, Tetraploid, Polyploid.*

Hardening

The increase in resistance to frost in a plant tissue.

Hardiness

The capability of plant tissue to survive the formation of ice crystals within them. cf. *Frost resistance.*

Hardpan

A cemented, hardened layer in the soil, cemented by iron oxide, silica, organic matter, or some other substance. cf. *Claypan.*

Hard Seed

The condition of some seeds in which water absorption and germination do not occur although the environmental conditions are favorable, e.g., seeds of many legumes and trees.

Hardwood

The wood of a tree in the *Angiosperms* (q. v.), e.g., the oak, in contrast to *Softwood* (q. v.).

Haustorium

A specialized organ of certain parasitic plants which penetrates the *Host* and absorbs food from its tissues, e.g., special *hyphal* branch, a specialized structure in dodder.

Hay

The dried, mowed stems and leaves, often including also flowers or fruits and seeds, of grasses and forage legumes such as alfalfa and clovers. cf. *Fodder, Stover.*

Headquarters

The main center or centers of activity of an animal, e.g.,

the larvae of the longicorn beetle in dying or dead wood. cf. *Habitat, Niche.*

Heat Budget

The amount of heat expressed in gram calories that is necessary to raise the temperature of a body of water from the winter to summer temperatures.

Heath

A community usually occurring in cool climates, often dry, usually without trees and uncultivated, characterized by low shrubby plants mostly in the family Ericaceae. cf. *Moor, Bog, Marsh, Swamp.*

Heaving

The partial raising of plants out of the ground resulting from the freezing and thawing of the soil during the winter, often breaking the roots.

Hedgerow

A strip of shrubs or small trees, often planted, enclosing a field or other area.

Heel-in

Placing young plants such as fruit trees in a temporary trench with soil over the roots to protect them from drying until they are permanently planted.

Hekistotherm

A plant growing in a cold region, which requires less heat than other kinds of plants, growing where the average temperature for the warmest month of the year is less than 50°F., e.g., mosses, lichens, sedges, etc., in the *Arctic.* cf. *Megatherm, Mesotherm, Microtherm.*

Heliophyllous

Refers to organisms that grow best in full sunlight.

Heliophyte

A *Heliophyllous* organism.

Heliophobous
Refers to organisms which grow best in the shade. Syn. *Sciophyllous, Shade plant.*

Heliotropic
See *Phototropic.*

Helminth
A worm, usually parasitic.

Helophyte
One of Raunkiaer's *Life-form classes,* consisting of marsh plants whose *Perennating* parts are in the soil under water, e.g., arrow-head. cf. *Geophyte, Hydrophyte.*

Helotism
A term that has been used for the interaction of two organisms such as an alga and fungus in a lichen. See *Mutualism, Parasitism, Symbiosis.*

Hematypic
Refers to reef-building corals which contain zooxanthellae and algal *Symbionts.*

Hemerocology
The ecology of land that is modified by man such as gardens, parks, fields. cf. *Culture community.*

Hemerophyte
A plant introduced into an area by man. cf. *Culture community, Exotic, Ruderal.*

Hemicryptophyte
One of Raunkiaer's *Life-form classes* in which the buds or *Perennating* parts of plants surviving unfavorable periods are located at the soil surface. cf. *Cryptophyte, Geophyte, Chamaephyte.*

Hemiepiphyte
A plant that is *Epiphytic* (q. v.) for only part of its lifetime.

Hemiparasite

An organism which is intermediate between an *Epiphyte* (q. v.) and a *Parasite* (q. v.).

Hemisaprophyte

A plant which is intermediate between a *Saprophyte* (q. v.) and an *Autotrophic* (q. v.) organism.

Hepatic

Refers to the liver or to a *Liverwort,* (q. v.).

Herb

A plant with one or more stems that die back to the ground each year; grasses and *Forbs* (q. v.) as distinct from shrubs and trees.

Herbaceous

Refers to plants with characteristics of an herb.

Herbage

Herbs in a collective sense, and any other plant material used as forage by animals, especially *Pasturage.*

Herbarium

A collection of preserved, classified plants.

Herbicide

A chemical substance used for killing plants particularly weeds, e.g., 2,4-D. cf. *Insecticide.*

Herbivore

An organism that eats plants, e.g., rabbit, sheep.

Herbivorous

Refers to a herbivore.

Herbosa

Vegetation consisting of non-woody plants.

Herd

A group of animals, especially cattle or big game. cf. *Band.*

Herding, Close

Handling a band of sheep or goats in a compact group and restricting the spread of the animals while grazing.

Herding, Open

Handling a band of sheep or goats so the individuals are allowed to spread freely while grazing.

Heredity

The transmission of characters or directions from parents to offspring, or the sum total of such characters.

Hermaphrodite

A plant with stamens and pistil in the same flower or an animal that produces both male and female *Gametes,* e.g., the rose, earthworm. cf. *Dioecious, Gynandromorph.*

Herpesian

Refers to reptiles or amphibians.

Herpetology

The branch of zoology concerning reptiles and amphibians.

Heteroecious

Refers to organisms having stages of its life-cycle on different hosts, e.g., wheat rust which attacks wheat and the barberry.

Heterogamy

(1) The production of unlike *Gametes* (egg and sperm), cf. *Isogamy.* (2) *Alternation of generations* (q. v.).

Heterogenesis

See *Alternation of generations.*

Heteromorphic

Refers to *Alternation of generations* (q. v.) in which forms in different stages are unlike.

Heterophagous

Refers to animals which eat several kinds of food. cf. *Omnivorous.*

Heterophyte

A *Saprophyte* or *Parasite* (q. v.). See *Heterotrophic, Autotrophic, Holophytic.*

Heteroploid

Refers to a population comprising *Aneuploid, Diploid,* and *Euploid* members.

Heterosis

The increase in vigor or the growth of offspring resulting from the crossing of genetically different parents. syn. *Hybrid vigor.*

Heterotrophic

Refers to an organism in which complex materials, especially organic foods, are the chief source of nutrition, in contrast to *Autotrophic* (q. v.) organisms, e.g., animals, *Parasites, Saprophytes,* cf. *Holozoic, Mixotrophic.*

Heterotropous

Refers to an animal which may wander into a community and play an important or unimportant part while present, but eventually dies if it does not move into a more favorable environment. cf. *Index species, Tychocoen.*

Heterozygous

Refers to an organism that originated from the fusion of gametes containing unlike Genomes (q. v.). cf. *Homozygous.*

Hibernaculum

(1) The habitat *Niche* where certain animals overwinter. (2) An overwintering bud of an aquatic plant.

Hibernal

Refers to winter, especially the winter season in contrast to the *Vernal, Estival,* and other seasons. cf. *Aspection.*

Hibernation

A state of *Dormancy* especially during winter. cf. *Estivation.*

Hiemal

See *Hibernal.*

Hiemilignosa

Monsoon forest which merges into savanna or park land in a tropical-continental climate; woody plants lose their small xerophytic leaves during the hot and dry summers.

Hierarchy

A social rank-order of animals formed through passive submission, direct combat, or threat.

High Grass

A class of grasses, 6 to 8 feet high or more, e.g., *Panicum virgatum.* cf. *Medium-height grass, Short grass.*

Histology

The division of biology that deals with microscopic structures or tissues of organisms.

H Layer

A layer of completely decomposed litter, unrecognizable as to origin, on the surface of the mineral soil. See *A horizon, F layer, L layer.*

Hoar Frost

A deposit of ice crystals formed directly from water vapor in the air. cf. *Rime.*

Hochmoor

A moor that rises from the edge towards the middle, so

that the upper surface is convex in section, caused by the growth of Sphagnum.

Hogback

A ridge with a narrow summit and steep slopes.

Hohenheim System

A system of grazing followed by resting small pastures or paddocks for short periods in rotation.

Holarctic

Refers to the combined *Palearctic* (q. v.) and *Nearctic* (q. v.) regions of the faunal realm *Megagea* (Arctogea) (q. v.).

Holard

A term used occasionally to designate the total water content of the soil.

Holism

The doctrine that life in all its forms and the inorganic environment form an interacting, integrated system. cf. *Ecosystem.*

Holocoen

Equivalent to *Ecosystem* (q. v.).

Holocoenotic

Refers to the joint action and interaction of a number of environmental factors upon organisms.

Holoparasite

An organism that is wholly parasitic, e.g., wheat rust.

Holophytic

Refers to an organism that utilizes light as the primary source of energy, e.g., green plants, purple bacteria. cf. *Autotrophic, Chemotrophic, Heterotrophic, Holozoic.*

Holoplankton

An organism that is in the *Plankton* (q. v.) during its entire life-cycle, e.g., *Copepod.* cf. *Meroplankton.*

Holozoic

Refers to a *heterotrophic* animal that ingests solid food, digesting it internally.

Homeostasis

The maintenance of constancy or a high degree of uniformity in functions of an organism or interactions of individuals in a population or community under changing conditions, because of the capabilities of organisms to make adjustments. Perceptual homeostasis is the state of maximum predictability and control of environmental stimuli by an organism. cf. *Steady state, Stabilization.*

Homeotherm

See *Homoiotherm.*

Home Range

The area around an animal's established home which is traversed in its normal activities. cf. *Territory.*

Homesite

The location of the nest or resting place that is in regular use by an animal.

Homing

The reaction of an animal to return to a given place after displacement, e.g., homing pigeons.

Homogeneity

Refers to the regularity in the distribution and abundance of the species in a community or area. cf. *Frequency, Homoteneity.*

Homoiotherm

An animal which is able to maintain the temperature of the body at an approximate constant level independent of the surrounding medium, "warm-blooded," e.g., birds, mammals. cf. *Poikilotherm.*

Homologous

Refers to structures of organisms that possess the same phylogenetic origin, e.g., wings of bats and forelimbs of a rabbit.

Homologous Chromosomes

The two *Chromosomes* (q. v.) occurring in each pair in *Diploids,* each derived from a separate parent.

Homologue

A pair of *Homologous chromosomes,* (q. v.).

Homoplastic

Refers to organisms which have organs resembling each other because of the evolution along similar paths, but the organs are not *Homologous,* e.g., *Lagomorphs* and *Rodents.*

Homoteneity (Homotony)

The *Homogeneity* (q. v.) of a *Vegetation* type in contrast to that of a *Stand.*

Homozygous

Refers to an organism resulting from the fusion of *Gametes* carrying the same genes. cf. *Heterozygous.*

Hook Order

A social order or rank in horned animals determined by the aggressive use of horns.

Hopkins Bioclimatic law

See *Bioclimatic law.*

Horizon (Soil)

A layer of soil approximately parallel to the soil surface, with distinct characteristics produced by soil-forming processes. See *A horizon, B horizon, C horizon.*

Hormone

A chemical substance produced in one part of an organism and usually transported to another part where it causes an effect. cf. *Auxin.*

Host

An organism that furnishes food, shelter, or other benefits to another organism of a different species. cf. *Parasite, Symbiont, Mutualism.*

Hudsonian Life Zone

One of *Merriam's Life zones,* includes the northern part of the *Boreal forest* and coniferous forest on mountains farther south; southern boundary delimited by the 57.2°F. isotherm for the six hottest weeks of the year.

Humidity, Absolute

The actual quantity of water vapor present in a given volume of air, usually expressed in grams per cubic meter.

Humidity, Relative

The ratio of the actual amount of water vapor present in a unit portion of the atmosphere to the quantity which would be present when saturated. cf. *Psychrometer, Hygrometer.*

Humification

The process of decay of organic material to *Humus.*

Humus

(1) Organic matter in a more or less stable advanced stage of decomposition, dark in color, with a high nitrogen content, a carbon-nitrogen ratio near 10:1, and other chemical and physical properties such as a high *Base exchange capacity,* water absorption, and swelling. (2) Residues in the soil of plants and animals that have undergone an appreciable degree of decomposition.

Humus, Raw

See *Mor.*

Hurricane

(1) A tropical cyclone, especially one in the West Indian region; with winds of hurricane force which blow around the central calm area or "eye" which is very low in atmospheric pressure. (2) The highest wind velocity on the *Beaufort scale* (q. v.), a wind greater than about 75 miles per hour.

Hybrid

(1) Genetic; an organism resulting from a cross between parents with different *Genotypes* (q. v.). (2) Taxonomic; a cross between parents of different *Taxa* (q. v.).

Hybridization

The crossing or breeding of unlike individuals to produce hybrids.

Hybrid Segregate

A form produced in the second or later generation after hybridization.

Hybrid Swarm

A population of organisms derived through hybridization, comprising various generations of hybrids and backcrosses; often varying greatly. cf. *Back-crossing.*

Hybrid Vigor

See *Heterosis.*

Hydathode

A pore or gland, usually in leaves, that exudes water. cf. *Guttation.*

Hydrarch

Refers to a *Succession* or *Sere* which begins in wet habitats such as a pond. cf. *Hydrosere, Xerarch.*

Hydraulic Equilibrium

The condition of absence of flow rate of water in soil,

when the pressure gradient force is equal and opposite to the gravity force.

Hydric

(1) Refers to, or containing hydrogen, e.g., hydric oxide. (2) Sometimes used wrongly in the sense of "wet" as a substitute for *Hydrophyte* (q. v.). See *Hygric.*

Hydrochore

A plant whose *Diaspores* (q. v.) are disseminated primarily by water, e.g., water lilies. cf. *Anemochore.*

Hydroclimograph

A graph in which monthly temperature data are plotted against data on salinity in the form of a polygon.

Hydrocole

An animal living in water. cf. *Hygrocole, Hydrophyte, Mesophyte, Xerocolous.*

Hydrogen-ion Concentration

The concentration of free hydrogen ions in a solution, commonly expressed as the logarithm of the reciprocal of the normality of free hydrogen ions in which pH 7.0 is neutral, values higher than 7.0 indicate alkalinity, below this acidity. cf. *Reaction.*

Hydrography

The study of natural bodies of water such as lakes, rivers, and seas, especially their physical characteristics in contrast to the biological qualities.

Hydrologic Cycle

The cycle of the movement of water from the atmosphere by precipitation to the earth and its return to the atmosphere by interception, evaporation, run-off, infiltration, percolation, storage, and transpiration.

Hydrology
The science dealing with water and snow, including their properties and distribution.

Hydromorphic
(1) Refers to soil occurring in poorly drained depressions in humid regions. cf. *Halomorphic.* (2) Refers to cellular features typical of *Hydrophytes.*

Hydrophilous
Refers to a plant that grows well in water or wet land.

Hydrophyte
A plant which grows wholly or partly immersed in water. cf. *Hygrophilous, Xerophilous, Mesophyte.*

Hydroponics
The growing of plants so that the roots are immersed in a water solution of nutrient salts or in some inert material such as vermiculite which is supplied with a nutrient solution.

Hydrosere
A collective term which includes all the stages in a *Succession* beginning in water. cf. *Hydrarch, Xerarch, Sere.*

Hydrosphere
The parts of the earth covered with water, including streams, lakes, oceans. cf. *Lithosphere, Biosphere.*

Hydrotropism
A growth response of plants to water as a stimulus.

Hyetograph
A form of self-recording *Rain gage.*

Hygric
Refers to a wet or moist condition of a habitat. cf. *Hygrophilous, Xerophilous.*

Hygrochase

A seed pod that opens in humid air and closes in dry air. cf. *Xerochase.*

Hygrocole

An animal living in a moist place. cf. *Hydrocole, Hygrophilous, Xerocolous.*

Hygrogram

A record made by a *Hygrograph.*

Hygrograph

A self-recording *Hygrometer.*

Hygrometer

An instrument for measuring the *Relative humidity* of the air.

Hygropetric

Refers to an organism that inhabits steep and wet rock surfaces.

Hygrophilous

A plant which grows in moist or wet places. cf. *Hygrocole, Xerophilous, Mesophyte.*

Hygroscopic Coefficient

The moisture in percentage of oven-dry (100-110° C.) weight that a soil will absorb in a nearly saturated atmosphere (relative humidity of 98 per cent at 25°C.).

Hygroscopic Water

Water held so firmly by the attraction of soil particles that it can be removed only by heating above 100°C. It is not available to plants.

Hygrothermograph

An instrument that makes a simultaneous record of both relative humidity and temperature. cf. *Hygrograph, Thermograph.*

Hymenoptera
An order of insects which includes the bees, ants, wasps, and ichneumon flies.

Hyperdispersion
A pattern of distribution of individuals of a species in an area which is characterized by clumping, or the occurrence of denser aggregations in some spots than in others. Syn. *Contagious distribution, Over-dispersion.* cf. *Hypodispersion, Normal dispersion.*

Hyperparasite
An organism that is parasitic upon another *Parasite.*

Hyperplasy (Hyperplasia)
An abnormal increase in the number of cells in an organism, e.g., certain plant galls, tumors in animals.

Hypertrophy
Usually used to mean an abnormal enlargement, in respect to an organism it may include both enlargement of cells and *Hyperplasy* (q. v.).

Hypha
A filament or thread-like structure of a *Fungus.*

Hyphal
Refers to a *Hypha.*

Hypocotyl
The portion of a seed or a seedling between the attachment of the *Cotyledons* and the *Radicle.*

Hypodermis
The layer of cells adjacent to the epidermis in certain plants, in some species the cells are thick-walled, in others used for water storage.

Hypodispersion
A pattern of distribution of individuals of a species in

an area which shows more even spacing than can be expected by chance. Syn. *Under-dispersion.* cf. *Hyperdispersion, Normal dispersion.*

Hypogeal

Refers to *Cotyledons* which remain underground after seed germination, e.g., the pea seed. cf. *Epigeal.*

Hypolimnion

The non-circulating body of water in lakes below the *Thermocline.* cf. *Epilimnion.*

Hyponasty

The more rapid growth of the under side of an organ such as a leaf than of the upper side.

Hypoplasy (Hyperplasis)

The reduction in the number of cells in an organ, e.g., in certain plant galls.

Hypsometer

An instrument for measuring the height of an object, especially a tree.

Hythergraph

A graphic presentation of climatic conditions in which one coordinate is the mean monthly temperature and the other is the mean monthly precipitation. cf. *Climagraph.*

I

Ice Age

A glacial epoch in which glaciers and ice sheets occupied large areas of continents, as occurred in the *Quaternary* period.

Ichthyology

The branch of zoology which deals with fishes.

Ideograph

A pictorial diagram of an object in which relationships are shown between measurements of various characteristics.

Idiobiology

The study of the individual organism. cf. *Autecology, Synecology, Individual ecology.*

Illuminance

Syn. *Light intensity.* A unit of illuminance is the lumen per square foot, equivalent to the foot-candle when *Light intensity* is used. cf. *Illumination value.*

Illumination Value

The illumination capacity or brightness of light as perceived by the human eye; on a clear summer day it is 8000 to 10,000 foot-candles at noon. cf. *Irradiance.*

Illuviation

The accumulation of material in a soil horizon by precipitation from solution or from suspension from a layer above. cf. *Eluviation, B horizon.*

Imago

The adult, sexually mature stage of an insect.

Immature Soil

A soil in which development is not complete because of insufficient time since deposition or exposure to the action of the physical environment and organisms on the *Parent material* (q. v.).

Immigrant

A plant or an animal of a species which has recently invaded an area, whose role is still uncertain. cf. *Emigration.*

Immigration

The *Migration* (q. v.) of an organism into an area where it did not occur previously. cf. *Emigration.*

Immunity

The capability of an organism to resist infection by a *Parasite* or one of its products.

Imperfect Flower

A flower lacking pistils or stamens.

Impervious Soil

Refers to soil or a certain layer in which the penetration of water, and usually air and roots as well, occurs slowly or not at all.

Impoundment

An artificial lake or pond.

Imprinting
A form of rapid and stable learning in a young animal when it is exposed to a meaningful stimulus.

Inalpage
The residence of flocks and herds upon *Alpages* (q. v.).

Inbreeding
The mating of closely related organisms. cf. *Outbreeding.*

Incompatibility
A failure or partial failure of some process which results in lack of *Fertilization.* See *Self-incompatibility.*

Increaser
A plant already present in a community or area which increases in abundance under overgrazing. cf. *Decreaser.*

Increment
The increase in the *Basal area* (q. v.), diameter, height, volume, quality, or value of a tree or a *Stand.*

Increment Borer
An instrument used for securing from the trunk of a tree a core which shows the *Growth rings.*

Indehiscent
Refers to a structure, especially fruits of plants, that do not break open when ripe. cf. *Dehiscent.*

Index, Frequency
See *Frequency index.*

Index of Similarity
The ratio of the number of species found in common in two communities to the total number of species that are present in both, cf. *Coefficient of community.*

Index Species
An organism that is so well adapted to its habitat it

seldom occurs elsewhere, therefore useful in characterizing the environmental conditions as a living label. cf. *Autochthonous.*

Indicator

An organism, species, or community which indicates the presence of certain environmental conditions. cf. *Exclusive species.*

Indifferent Species

A species occurring in many different communities, a *Companion species* (q. v.). See *Fidelity.*

Indigen

An *indigenous* species. cf. *Cultigen.*

Indigenous

Refers to an organism that is native, not introduced, in an area. cf. *Endemic, Exotic.*

Individual Ecology

See *Autecology, Idiobiology, Synecology.*

Infection

The establishment of a *Parasite* upon an organism.

Infest

The establishment of an organism in numbers as parasites upon another plant or animal, e.g., aphids infesting a rose bush.

Infiltration

The penetration of water into soil or other material. cf. *Percolation.*

Infiltration Capacity

The maximum rate of Infiltration under a given set of conditions.

Infiltrometer

A mechanism for measuring *Infiltration* into the soil in which water is applied by sprinkling or flooding.

Inflorescence

The flower-cluster in plants, including the flowers, bracts, and stems.

Influent

(1) An organism which has important interactions (*Reactions, Coactions*) within a community, but is not a dominant. (2) The flow of water from a channel into subterranean storage. cf. *Effluent.*

Infraneuston

Animals which are supported by the lower surface of the film of water on lakes and ponds, e.g., mosquito larvae. cf. *Neuston, Supraneuston.*

Infusoria

Used formerly in the sense of all microscopic organisms occurring in infusions of organic matter; now used chiefly for ciliated *Protozoans* (Ciliophora).

Ingestive Behaviour

The actions of an organism when eating or drinking.

Inheritance

See *Heredity.*

Inheritance of Acquired Characters

The outmoded theory of evolution that modifications produced during the lifetime of an individual, because of use or disuse, new needs, or because of the direct action of the environment are inherited by the offspring and are cumulative with time.

Initial Cause of Succession

The process or agent which produces denuded or partly

denuded areas on which ecologic *Succession* is initiated, e.g., *Erosion, Deposition.*

Innate Releasing Mechanism
A device (postulated) in the nervous system which initiates a certain reaction when an animal receives a particular stimulus.

Inoculate
To introduce a microorganism, virus, serum, etc., into an organism.

Insecticide
A material used for destroying insects, e.g., DDT, rotenone. cf. *Herbicide, Fungicide.*

Insectivore
An animal in the order Insectivora, a primitive *Insectivorous* group in the class Mammalia, e.g., mole, shrew.

Insectivorous
Refers to an organism that eats insects.

Insect Vector
An insect by which a disease-producing organism or a parasite is disseminated, e.g., aphids and leafhoppers transmitting plant diseases.

Insolation
Solar radiation received by the earth or other planets from the sun, or exposure to rays of the sun.

Instar
A form in the larval development of insects between two moults.

Instinct
An inherited and adapted system of co-ordination within the nervous system as a whole, which when activated finds expression in behaviour culminating in a fixed action pattern.

Interaction

An interrelationship between organisms, between organisms and the environment, or between environmental factors, e.g., *Competition,* grazing, wilting of a plant, *Relative humidity* decreased by heat. See *Coaction, Reaction.*

Intercellular

Refers to the position between cells of an organism, e.g., air-spaces between cells in leaves.

Interception

The process by which precipitation is retained by leaves, branches, and other organs of plants before the moisture reaches the ground.

Interference

The ability of a species to damage another either directly by attacking its individuals or indirectly by harming its resources or blocking access to them. cf. *Competition, Exploitation.*

Interfluve

A ridge between river valleys.

Internal Drainage (Soil)

The quality of a soil that permits downward flow of excess water through it, determined by the texture, structure, depth to the *Water table,* etc. cf. *Gravitational water.*

Internal Environment

The conditions within an organism or cell that influence its processes, e.g., the oxygen content in body fluids in animals or in air-spaces in plants.

Interspecific

Refers to relations or conditions between species. cf. *Intraspecific.*

Interspecific Association

See *Association, interspecific.*

Interspersion

The irregular occurrence of plant communities and species which provide cover for animals within a limited area. cf. *Mosaic.*

Intertidal Zone

See *Tidal zone.*

Intracellular

Refers to the location or position of a substance or structure within a cell.

Intraclinal

Refers to the presence of organisms such as *Ecotypes* (q. v.) within a *Cline* (q. v.).

Intradiel

Refers to the period of a single 24-hour day.

Intraneous

The presence of individuals of a species toward the center of its entire area of distribution. cf. *Extraneous.*

Intraspecific

Refers to relations or conditions between individuals within a species. cf. *Interspecific.*

Intrazonal Soil

A group of soils having characteristics caused by the preponderant influence of local relief or parent material over the normal influences of the prevailing climate and vegetation. cf. *Zonal soil.*

Intrinsic Cycle

See *Cycle.*

Introgression

See *Introgressive hybridization.*

Introgressive Hybridization

The infiltration of genes of one species by the intermediacy of *Hybrids* into another species, resulting in the genetic modification of the latter.

Invasion

The *Migration* (q. v.) and *Establishment* (q. v.) of an organism in a new location.

Inverse Stratification

The condition in which the water just beneath the ice in a body of water is near the freezing temperature and within a short distance below shows a rapid rise to 3°C., and further below a gradual increase to 4°C., or to the maximum temperature of the lake or pond. cf. *Thermal stratification.*

Inversion, Temperature

An increase in air temperature with an increase of altitude, instead of the normal decrease.

Invertebrate

An animal lacking a spinal column, e.g., insects.

In Vitro

Refers to experiments on cells, etc., which are carried on when they are separated from the living organisms, e.g., tissue cultures. cf. *In vivo.*

In Vivo

Refers to location within the living system. cf. *In vitro.*

Involucre

A number of closely associated bracts subtending a flower or flower cluster.

Involution

(1) The diminution in the size of an organ. cf. *Hyperplasy, Hypertrophy.* (2) The formation of abnormal yeasts, bacteria. etc.

Ion Exchange

The replacement of one kind of ion by another, e.g., hydrogen ions replacing calcium ions in certain soil solutions. See *Exchange capacity*.

Ionization

The process of ion formation.

Ionizing Radiation

Radiation that takes electrons from atoms and attaches them to other atoms, e.g., *Alpha, Beta, Gamma radiation,* (q. v.).

Irradiance

The receipt of radiant energy per unit area per unit of time. On a clear summer day solar radiant energy equals 1.2 to 1.5 gram-calories per square centimeter per minute at noon. cf. *Illuminance, Light intensity.*

Irradiation

The exposure of an object to radiation such as sunlight, *Ionizing radiation,* etc.

Irritability

The characteristic capability of an organism to respond to a stimulus, e.g., a plant growing towards the light.

Irruption

An abrupt, irregular increase in population number or size.

Isobar

A line drawn on a map or chart connecting places of equal barometric pressure.

Isobath

A line drawn on a map connecting points of equal depth on the bottom of the sea.

Isobiochore

A line drawn on a map connecting regions that possess similar *Biological spectra* (q. v.).

Isocies

A group of *Synusiae* (q. v.) that show *Physiognomic* similarity.

Isoflor

A line drawn on a map connecting regions possessing an equal number of species within a genus or a family.

Isogamy

The production of similar *Gametes* (q. v.), occurs in certain algae (*Ulothrix*), fungi, and protozoa, cf. *Heterogamy*.

Isogenous

Refers to organisms that occur in the same region.

Isogram

See *Isotherm*.

Isohaline

The line or layer within a body of water which has the same *Salinity* at a certain time or the same mean salinity over a certain period.

Isohel

A line drawn on a map connecting places with equal duration of sunshine.

Isohyet

A line drawn on a map connecting places with equal quantity of rainfall.

Isolation

The separation of populations from other populations of the same species by geographic, ecologic, climatic, physiologic, or other barriers. cf. *Natural selection*.

Isolation Transect

An *Exclosure* (q. v.) which is divided into plots, one of which is opened to grazing each year, and another plot which has been grazed is added.

Isonome

A line drawn on a chart connecting areas of a community that show equal *Frequency indices* of a species.

Isophene

A line drawn on a chart connecting areas where events in the life history (e.g., egg-laying, flowering) of an organism occur at the same time. cf. *Aspection.*

Isopleth

A line drawn on a map or chart connecting places having the same value of a certain factor. cf. *Isohyet, Isotherm.*

Isopod

An animal in the order Isopoda, class Crustacea, e.g., woodlice, pillbug.

Isostasy

The state of general equilibrium between the upland and lowland areas of the earth, with indications that the rock materials under the oceans are heavier than those under continental protuberances. cf. *Tectonic.*

Isotherm

A line drawn on a map or chart connecting places with the same temperature at a particular time or for a certain period.

Isotopes

Forms or atoms of the same element that differ in atomic weight and in the constitution of the atomic nucleus. Some elements in nature such as radium and uranium have *Radio isotopes* (q. v.).

Itograph

An instrument arranged at the entrance of a bird's nest for the automatic recording of the number and direction of visits made by the parents.

J

Jarovisation

See *Vernalization.*

Jordanon

A *Microspecies* (q. v.).

Jordan's Rule

The generalization that fishes living in waters of low temperatures tend to have more vertebrae than do those in warm waters.

Jurassic

A geological period in the Mesozoic era which began about 165 million years ago and lasted about 30 million.

K

Kame

A short ridge or mound of sand or gravel desposited by a stream under a glacier. cf. *Esker.*

Karst

Refers to a limestone region with a dry, barren surface from which most or all of the drainage is through underground channels.

Kar Herbage

An aggregation of tall herbs growing in fertile soil in hollows high in mountainous regions.

Karroo

An open vegetation type in South Africa consisting of succulent and sclerophyllous shrubs, where the precipitation amounts to 3 to 14 inches annually, but which falls mostly in the summer.

Karyokinesis

See *Mitosis.*

Karyotype

The gross appearance, i.e., the size, number, and shape, of the set of *Somatic chromosomes.*

Key-Industry (Animals)

Herbivorous animals which are so numerous that a large number of other animals are dependent upon them for food (e.g., *Copepods*. cf. *Food-chain, Pyramid-of-numbers.*

Key Areas

Critical areas of range land which represent range that is most likely to be overgrazed; used as criteria or indices of proper use of the range.

Key Species

Any species of plants which because of palatability, abundance, or other characteristics may be used in estimating degree of utilization, trend, or condition of the range. cf. *Decreasers.*

Krebs Cycle

The aerobic portion of *Respiration,* in which pyruvic acid is oxidized, usually to carbon dioxide and water as end products.

Kinesis

The behaviour of an animal resulting from unoriented reflex action of the entire animal.

Klendusity

The capability of an otherwise susceptible variety of a species to escape infection because of the way it grows, e.g., plants that mature early and thus escape late-season diseases.

Klinokinesis

The random turning movements of an organism which increase in rate as it nears an unfavorable environment. cf *Klinotaxis, Orthokinesis.*

Klinotaxis

A sudden movement away from an unfavorable environment, directed by the organism. cf. *Klinokinesis, Orthokinesis.*

Koprophagous

See *Coprophagous.*

Krotovinas

Irregular, tubular streaks within one soil horizon, consisting of material transported from another horizon; caused by filling of tunnels made by burrowing animals, especially rodents.

Krummholz

Scrubby, stunted growth-form of trees, often forming a characteristic zone at the limit of tree growth in mountains.

L

Lacustrine
Refers to a lake.

Lagg
See *Raised bog.*

Lagomorph
An animal in the order Lagomorpha, class Mammalia, e.g. rabbit.

Lamarckism
The doctrine regarding the inheritance of *Acquired characters* (q. v.) propounded by J. B. Lamarck.

Lambing Range
The area used by bands of sheep during the lambing season.

Land Bridge
A land connection between two bodies of land over which migration of organisms has occurred.

Land Capability
The suitability of land for use of some kind without damage.

Land-capability Class

One of the eight classes of land in the land-capability classification, ranging from (1) land that is very good for cultivation to (8) land that is not suitable for cultivation, grazing, or forestry.

Larva

The pre-adult, usually self-feeding, but not sexually reproducing form of an animal, passes through metamorphosis to the adult stage, e.g., caterpillar of a moth, tadpole of a frog.

Lasion

A *Periphyton* (q. v.) in which the organisms are associated in a more or less dense growth and are interdependent. cf. *Epiphyton.*

Laterite

A red, highly weathered soil characteristic of damp tropical regions such as equatorial Africa. cf. *Laterization.*

Laterization

Weathering which tends to produce *Laterite,* essentially, the removal of silica and consequent increase in alumina and iron oxide content, and a decrease in the *Base exchange capacity* of the soil. cf. *Podzolization.*

Laurilignosa

Laurel forests or subtropical rain forests, often with *Dicotyledonous* and *Gymnospermous* dominants. cf. *Lignosa.*

Layer

The horizontal part of a community in which the plants are of about the same height, e.g., tree layer, herb layer. Also applicable to depth in the soil. syn. *Stratum.* cf. *Layering, Synusia.*

Layerage

The propagation of plants by inducing formation of roots on stems that are attached to the plant.

ayering

The appearance of plants or plant parts, or their remains, in horizontal divisions. syn. *Stratification.*

Leaching

The removal by percolating water of soluble constituents from the soil or other material.

Leafmold

The lower layer of the A_0 *horizon,* lying on the mineral soil, consisting mostly of well-decomposed, finely-divided organic material.

Leaf-size Classes

The arbitrary groups of leaves based on the area of blades, as proposed by Raunkiaer, in square mm: *Leptophyll* 25, *Nanophyll* 225, *Microphyll* 2025, *Mesophyll* 18,225, *Macrophyll* 164,025, *Megaphyll* larger than 164,025.

Legume

(1) A plant belonging to the family Leguminosae, e.g., pea, alfalfa. (2) The fruit of Leguminosae.

Leguminous

Refers to the pea family, Leguminosae.

Lemming

One of the small rodents in genus *Lemmus* or *Dicrostonyx,* order Rodentia, of circumpolar distribution.

Length-of-day

See *Photoperiodism.*

Lentic

Refers to the standing-water series; lakes, ponds, swamps. cf. *Lotic.*

Lenticel

A pore on the surface of woody stems or roots, filled with loosely arranged cells that permit diffusion of gases between the atmosphere and the interior of the plant.

Lepidopteron

An insect in the order Lepidoptera (e.g.) moth.

Leptophyll

See *Leaf-size classes.*

Lethal Gene

A *gene* (q. v.) that causes death of an organism.

Leucoplast

A colorless *Plastid* in which starch often forms, located in the *Cytoplasm* in plant cells.

Level Terrace

A terrace that strictly follows the contour, in contrast to the *Graded terrace* (q. v.).

Ley

An English term for land that is temporarily under grass, legumes, or mixtures of these.

Liana (Liane)

A climbing or twining plant.

Lichen

A *Symbiotic* association or relationship of an alga and a fungus, which forms crustose, foliose, or fruticose bodies.

Liebig's Law of the Minimum

The generalization that states the growth and reproduction of an organism is dependent on the nutrient substance, such as nitrogen, oxygen, carbon dioxide, that is available in minimum quantity.

Life Belt

A vertical subdivision of plant and animal life, determined largely by altitudinal influences, part of a *Biotic province* (q. v.).

Life Cycle

The phases, changes, or stages an organism passes through

from the fertilized egg to death of the mature plant or animal.

Life Expectancy

The average duration of life that a given individual is expected to live after having reached a certain age. cf. *Life table.*

Life-form

The characteristic form or appearance of a species at maturity, e.g., tree, herb, worm, fish. cf. *Growth form, Habitat form, Raunkiaer's life-form classification.*

Life-form Class

One of the groups in Raunkiaer's classification of life-forms, e.g., *Geophyte, Therophyte.*

Life-form Dominance

The condition in which several species of the same *Life-form* dominate a plant community. cf. *Dominance, ecologic.*

Life Span

The maximum duration of life of an individual of a species.

Life Table

A statistical tabulation presenting complete data on the mortality of a population. cf. *Ecological longevity, Life Expectancy.*

Life Zone

An altitudinal or latitudinal biotic region or belt with distinctive faunal and floral characteristics. cf. *Alleghenian, Hudsonian life zones.*

Light Intensity

See *Illumination value.*

Light Quality

The wave-length composition of light.

Lignin

A complex organic compound in the walls of certain cells, especially in woody tissue.

Lignification

The process of impregnating cell walls of a plant with lignin.

Lignosa

Woody vegetation.

Lime Requirement

The amount of standard ground limestone needed to change the upper 6.6-inch layer of an acre of acid soil to some lesser degree of acidity, usually stated in tons per acre.

Limiting Factor

The environmental influence by which the limit of toleration of an organism is first reached and which therefore acts as the immediate restriction to one or more of its functions or activities or in its geographic distribution.

Limnetic

Refers to the open water of a pond or lake. cf. *Benthic.*

Limnology

The branch of biology that deals with fresh waters and organisms in them.

Lincoln Index

The use of marked animals to estimate the size of a population.

Line-Intercept Method

The sampling of vegetation by recording the plants intercepted by a measured line placed close to the ground, or by vertical projection to the line. cf. *Transect.*

Line-plot Survey

The sampling of vegetation by means of plots of uniform size located at regular intervals along a line.

Line Transect

Sampling vegetation by recording kinds of plants or communities intercepted by a measured line. cf. *Line- intercept method, Transect.*

Linkage

The association of certain characters in such a way that they are inherited together, because the controlling genes are in the same *Chromosome*.

Linnaean

Refers to the work or the concepts of Carolus Linnaeus.

Linneon

A species according to the nomenclature of Linnaeus, a broad category, often containing variable forms.

Lister

An implement consisting of a double plow, in which the shares push the soil in opposite directions, forming a series of alternate ridges and furrows. The Basin lister has an attachment that forms low dams of soil across the furrows at intervals of 15 to 25 feet, so that basins are formed which can hold large amounts of water.

List Quadrat

A rectangular sample area in vegetation in which organisms are merely tabulated according to species.

Liter

According to the metric system the volume of pure water, free of air, at 760 mm. pressure and 4°C., equivalent to 1.057 U. S. liquid quart.

Lithophyte

A plant growing on a rock, e.g., many lichens and mosses.

Lithosere

All of the stages of a successional sequence that originated on rock. cf. *Succession, Xerosere, Hydrosere.*

Lithosol

A soil consisting mainly of partly weathered rock fragments or of nearly bare rock.

Lithosphere

The earth's crust, consisting of the surface soil lying upon the hard rock which is several miles thick. cf. *Hydrosphere, Biosphere.*

Litter

(1) The uppermost organic materials, partly or not at all decomposed, on the surface of the soil. cf. A_{oo} *horizon.* (2) The group of young born at one time by a *Multiparous* animal as a cat.

Littoral

Refers to the zone in a lake or a pond that extends from the shore to the depth at which plants are rooted. In the ocean the zone extends to about the depth to which tides, wave action, and light penetrate.

Liver-fluke

A *Fluke* (q. v.) parasitic on sheep, cattle, and other animals, causing liver-rot.

Liverwort

A plant in the class Hepaticae, phylum Bryophyta, usually growing in moist places, e.g., Marchantia.

Llano

A tropical *Savanna* (q. v.) or grassland north of forests of the Amazon River basin in South America.

L Layer

Used at times for the A_{oo} *horizon* (q. v.).

Llorano

A winter fog caused by the invasion of cold air during "northers" along the shores of the Gulf of California.

Loam

(1) A soil containing relatively equal amounts of sand and silt and a somewhat smaller proportion of clay. (2) Specifically, soil material containing 7 to 27 per cent clay, 28 to 50 per cent silt, and less than 52 per cent sand.

Local

Refers to a relatively small area, a few square miles as a maximum.

Localization

The behaviour of an animal where it becomes associated with a particular area.

Local Race

A group of individuals of a species with better genetic adaptation to a given environment than other groups. cf. *Ecotype*.

Lociation

A local variation of a *Climax* community, differing from it in the kinds of *Subdominants*. cf. *Faciation*.

Locies

Similar to *Lociation* but applies to a *Seral* community.

Loess

A deposit of relatively uniform, fine soil material, mostly *Silt*, presumably transported to its present position by wind.

Logged-over

See *Cut-over forest*.

Logistic Curve

A graph that represents the growth of an individual or a population, typically S-shaped.

Long-day Plant

A plant that blooms under long periods of light and short periods of darkness, e.g., red clover. cf. *Photoperiodism.*

Loss-on-ignition

The loss in weight of a soil (or other material), previously dried at 100°C., heated to redness in a crucible; often used to represent the organic content.

Lotic

Refers to running water as in a creek. cf. *Lentic, Rheology.*

Lower Austral Life Zone

See *Austral life zone.*

Lower Sonoran Life Zone

See *Sonoran life zone.*

Luminescence

The emission of light that is not caused by high temperature. cf. *Bioluminescence.*

Lunar Periodicity

The correlation of activities of certain organisms with periods of the moon, e.g., *Bioluminescence* of the Bermuda fireworm at the time of full moon.

Lycopod

A plant in the subphylum Lycopsida (club-mosses) in the phylum Tracheophyta, e.g., *Selaginella.*

Lysimeter

An apparatus used to collect and measure the amount

of water that percolates through a quantity of soil and for measuring the amount of *Leaching.*

Lysin

A substance that causes bacteria, blood corpuscles, and other organic bodies to dissolve.

M

Macchia

Vegetation consisting of dense evergreen brush (shrubs and small trees) in the Mediterranean region, denser than *Garique* (q. v.), similar to *Chaparral* (q. v.). Syn. *Maquis.*

Macronutrient (Macrometabolic Element)

An element or a compound required by organisms in relatively large quantity, e.g., calcium by clams, phosphorus salts by clovers. cf. *Micronutrient.*

Macrophyll

See *Leaf-size class.*

Macrophytic

Refers to large aquatic plants, e.g., kelps, water cress

Macropterous

Refers to an animal with unusually large wings or fins. cf. *Micropterous.*

Macrospecies

See *Linneon.*

Maestro

A northwesterly wind in the central Mediterranean region.

Mafic

Refers to ferromagnesian minerals in a rock.

Malacology

The division of zoology that deals with mollusks.

Mallee

Scrub vegetation composed largely of various species of *Eucalyptus*, about 2 to 10 meters high, in dry, subtropical parts of southwest and southeast Australia.

Malthusian

Refers to the doctrine of T. R. Malthus that organisms tend to increase in geometrical progression while the food supply increases in arithmetical progression, so that the increase in the size of a population tends to be at a more rapid rate than the increase in available food. cf. *Competition*.

Mammal

An animal in the class Mammalia, subphylum Vertebrata, e.g., rabbit, deer.

Mangrove

A type of vegetation that is worldwide on tropical and subtropical saline, tidal mud flats, consisting usually of low trees or shrubs in genera *Rhizopora, Avicennia,* and *Sonneratia*.

Maquis

See *Macchia*.

Marl

A deposit of chiefly calcium carbonate, mixed with clay or other material, and formed chiefly in fresh water lakes by organisms such as *Chara*.

Marsh

A *Swamp* in which grasses, sedges, cattails, or rushes form the dominant vegetation. cf. *Bog. Moor*.

Marsh Gas
See *Methane.*

Marsupial
An animal in the subclass Marsupalia, class Mammalia, e.g., kangaroo, opossum.

Massif
A principal, relatively uniform mountainous mass with peaks on top.

Mass Selection
The choosing of individuals that possess a certain characteristic in common from a population such as a corn field, and then bulking the seeds or propagules for later planting.

Mast
Nuts such as acorns, beechnuts, and others, in a collective sense, especially when the nuts are used as food for animals.

Mature Soil
A soil that is in good adjustment with environmental conditions, in many regions with well-developed horizons. cf. *Podzol, Chernozem.*

Meadow
A grassland, usually in a low, moist area; often mowed for hay. cf. *Pasture, Range.*

Mean Sample Tree
A tree selected for its representative form and that is average in diameter, height, and volume of the other individuals of the species in a stand.

Mechanical Analysis
A laboratory procedure for determining the percentages of clay, silt, and sand in a sample of soil.

Mediterranean Climate
The climatic conditions that prevail in lands bordering

the Mediterranean Ocean, characterized by hot, dry summers and cool, rainy winters.

Medium-height Grass

In the classification of grasses according to height, the class that ranges from 2 to 5 feet in height, includes *Mid-grasses* (q. v.). cf. *Highgrass, Shortgrass.*

Megagea Realm

One of the three classes of the earth's fauna, which includes the Ethiopian, Oriental, Palearctic, and Nearctic regions. cf. *Neogea, Notogea.*

Megaphanerophyte

A group of plants in *Raunkiaer's life-form classification* which includes trees, lianas, and epiphytes over 30 meters high.

Megaphyll

See *Leaf-size classes.*

Megaspore

The larger of two kinds of spores produced by plants such as *Selaginella* and the *Spermatophytes* (q. v.). cf. *Microspore.*

Megatherm

An organism that requires continuously high temperatures, and according to some usage abundant moisture, e.g. sugarcane. cf. *Mesotherm, Microtherm, Hekistotherm.*

Meiosis

The two successive divisions of the nucleus in which the *Chromosome* number is halved, from the *Diploid* (q. v.) to the *Haploid* (q. v.) number.

Melanism

The unusual development of a dark pigment in an organism.

Melanocyte

Cells in certain animals that contain black pigment, melanin, as in the chameleon. The contraction of the cells makes the animal appear light in color, expansion makes it appear dark.

Melliphagous

Refers to an organism that feeds on honey.

Mendelian

Refers to *Mendel's laws* (q. v.).

Mendelian Population

A group of individuals of a species that share in a common *Gene* pool through reproduction. cf. *Species, Syngameon.*

Mendelism

The knowledge of inheritance according to *Mendel's laws.*

Mendel's Laws

The rules according to which characteristics of organisms are inherited as stated by Gregor Mendel, such as characters or factors (genes) act as units, dominance and recessiveness of characters, the segregation of *Alleles* during meiosis, and the independent assortment of alleles in each *Gamete.*

Mercator's Projection

A method of mapping in which the parallels of latitude are drawn as straight lines of the same length as the equator.

Meristem

A tissue in plants that is concerned with division to form new cells, located in various places such as root tips, stem tips, and buds. cf. *Cambium.*

Meroplankton

An organism that is in the *Plankton* (q. v.) during part of its life cycle. cf. *Holoplankton.*

Merriam's Life Zones

A series of belts or *Life zones* (q. v.) based originally on criteria of temperature according to C. Hart Merriam. See *Alleghanian, Carolinian, Hudsonian, Sonoran, Transitional, Tropical Life zones.*

Mesa

A flat or nearly flat table land with steep sides.

Mesarch

Refers to a successional series that begins in a moderately moist habitat. cf. *Hydrarch, Xerarch.*

Meseta

The greatly eroded, broad plateau in the interior part of Spain, crossed by a few mountain ridges.

Mesic

Refers to environmental conditions that are medium in moisture supply. cf. *Mesophytic, Hygric, Xeric.*

Mesolimnion

See *Thermocline.*

Mesophanerophyte

One of the groups of plants in *Raunkiaer's life form classification,* consisting of trees, lianas, and epiphytes, 8 to 30 meters tall.

Mesophyll

The palisade and sponge cells between the upper and lower epidermises in a leaf.

Mesophyte

A plant that grows in environmental conditions that are medium in moisture conditions, e. g., corn.

Mesophytic

Refers to a *Mesophyte.*

Mesosaprobic

Refers to an aquatic environment in which the oxygen content is considerably reduced and in which much decomposition of organic materials is taking place. cf. *Catarobic, Polysaprobic.*

Mesotherm

An organism that requires moderate warmth and moderate moisture, e. g., corn, hickory. cf. *Megatherm, Microtherm, Hekistotherm.*

Mesotrophic

Refers to a swamp supplied with a moderate amount of nutrients.

Mesozoic

One of the great geological eras, preceding the Cenozoic era, began about 205 million years ago and lasted about 130 million years; divided into the Triassic, Jurassic, and Cretaceous periods.

Metabolic Water

The water obtained from the chemical breakdown of foods by some organisms such as the clothes moth.

Metabolism

The sum total of chemical processes occurring within an organism or a portion of it, includes *Anabolism* and *Catabolism* (q.v.). Basal metabolism is the rate of expenditure of energy while an animal is at rest. cf. *Autotrophic, Heterotrophic.*

Metabolite

Any substance that plays a part, directly or indirectly, in *Metabolism.*

Metagenesis

See *Alternation of generations.*

Metalimnion

See *Thermocline*.

Metamorphosis

The change of an animal from one form to another in its postembryonic development, e. g., larva of an insect to a pupa.

Metaxenia

The differential effect of pollen from different varieties on the development of the fruit.

Metazoan

An animal in the group Metazoa which includes all multicellular animals as opposed to the unicellular *Protozoan* (q. v.).

Meteorograph

An apparatus for automatically recording simultaneously two or more meteorological elements.

Meteorology

The study that deals with physical processes occurring in the atmosphere such as precipitation, winds, and temperature.

Methane

CH_4, often called marsh gas, an odorless, inflammable gas, and explosive when mixed with air. Develops from decomposing organic matter in marshes and in coal-mines.

Micelle (Micella, Micell)

A particle composed of complex molecules that forms the units of structure in many organic substances such as cellulose and starch.

Microassociation

The abstract class or type of community in which similar *Microstands* are grouped, e. g., microstands of certain kinds of annual weeds that form a zone around carpenter-ant mounds in a shortgrass association or community-type.

Microbe

A *Microorganism* (q. v.).

Microclimate

The climatic conditions within a *Microhabitat* (q. v.). cf. *Local.*

Microcommunity

A small *Community* (q. v.) such as the plants and animals living in and on a decaying stump in a forest.

Microcosm

A miniature world, organisms plus the environmental conditions. cf. *Ecosystem.*

Microdissection

The procedure in which organisms or cells are studied through a microscope by means of a mechanically operated apparatus.

Microhabitat

A small *Habitat* (q. v.), e. g., a tree stump or a space between clumps of grass.

Micro-micron

A metric measure, one-millionth part of a *Micron* (q. v.).

Micro-millimeter

A metric measure, one-millionth part of a millimeter.

Micron

A metric measure, one thousandth part of a millimeter.

Micronutrient

A chemical substance required by an organism in very small quantity, e.g., boron by many plants. cf. *Essential element, Macronutrient, Vitamin, Deficiency disease.*

Microorganism
An organism that is microscopic in size, e. g., bacteria, protozoa.

Microphagous
Refers to an animal that feeds on particles that are very small in comparison to its own size, e.g., certain whales feeding on plankton.

Microphanerophyte
A group of plants in *Raunkiaer's life-form classification,* includes trees, shrubs, lianas, and epiphytes, two to eight meters tall.

Microphyll
See *Leaf-size classes.*

Microphyllous
Refers to a plant that has small leaves.

Micropterous
Refers to a fish with small fins or to an insect with small hind wings. cf. *Macropterous.*

Microrelief
Minor differences in topography such as small mounds or pits with differences in elevation of about three feet or less.

Microsere
A series of successful stages that occur within a microhabitat such as a tree-stump.

Microsome
A minute, RNA-rich particle in the cytoplasm of a cell, the center of protein synthesis.

Microspecies
A *Species* that is less inclusive than the *Linneon* (q. v.), similar to *Subspecies* (q. v.). Syn. *Jordanon.*

Microspore

The smaller of two kinds of spores produced by plants such as *Selaginella* and the *Spermatophytes,* e. g., pollen grains.

Microstand

A group of plants that occupies a *Microhabitat* (q. v.). cf. *Stand.*

Microtherm

An organism that can develop in cool and short summers, e. g., barley, spruce trees. cf. *Megatherm, Mesotherm, Hekistotherm.*

Microtome

An instrument for cutting very thin sections of tissue for microscopic study.

Mictium

A heterogeneous mixture of species such as occurs often in a transition zone between two kinds of stands.

Midgrass

A grass two to four feet tall, in contrast to a tallgrass which is five feet or more tall, e. g., *Koeleria cristata.* cf. *Medium-height grass, Shortgrass, Highgrass.*

Migrant

An organism that is undergoing *Migration* (q. v.).

Migration

(1) The movement of a plant or one or more of its parts, such as fruits, from one area to another. (2) The movement of an animal beyond its regularly occupied geographic location or *Home range* (q. v.). cf. *Emigration, Immigration, Invasion.*

Migrule

See *Diaspore.*

Milacre

An area one-thousandth part of an acre, containing 43.56 square feet, often used as a plot 6.6 feet square.

Millibar

A unit of atmospheric pressure, 1000 millibars represents a pressure of about 29.53 inches (750.1 mm.) of mercury.

Millicurie

One-thousandth part of a *Curie* (q. v.).

Milligram

One-thousandth part of a gram.

Milliliter

One-thousandth part of a liter.

Millimeter

One-thousandth part of a meter, 0.0394 inch.

Millimicron (mu)

One-thousandth part of a *Micron* (q. v.).

Mima-type Microrelief

A type of microrelief characterized by low mounds or soil pimples, named after the mounds in Mima prairie in western Washington.

Mimesis

The kind of behaviour in which like elicits like, involving hereditary patterns.

Mimetic

Refers to mimicking behaviour. cf. *Allelomimetic.*

Mimicry

(1) Batesian: The kind of behaviour in which an edible species escapes death by its close resemblance in appearance to an inedible species. (2) Mullerian: The kind of be-

havior in which both species are inedible but are similar in appearance, so avoidance learned by predators in tasting one is extended to the other. The term mimicry is often restricted to the former.

Mineralization

The decomposition of organic substances to mineral forms, e. g., proteins to nitrates, phosphates, etc.

Mineral Soil

Soil composed mainly of inorganic materials and with only a relatively low amount of organic material.

Minimal Area

The smallest area on which a community develops its *Characteristic species-combination* (q. v.).

Minimalraum, Minimiareal

See *Minimal area.*

Minimum, Law of

See *Liebig's law of the minimum.*

Minimum Quadrat Area

For a given number of samples in a stand the size of quadrat in which the *Species-area curve* (q. v.) becomes nearly horizontal, and the use of a larger size to secure greater accuracy is not justified by the time and labor involved.

Minimum Quadrat Number

For a given size of quadrat used to sample a stand the number of quadrats at which the number of species-number of quadrats curve becomes nearly horizontal, and the use of more quadrats to secure greater accuracy is not justified by the time and labor involved. See *Species-number curve.*

Minor Element

See *Essential element, Micronutrient.*

Miocene

A geological epoch which began about 28 million years ago and lasted about 16 million years, in the Tertiary period of the Cenozoic era.

Mississippian Period

A geological period in the *Paleozoic era* (q. v.), which began about 280 million years ago and lasted for about 25 million years.

Mistral

A cold, northerly wind along the northwest Mediterranean coast, especially during winter.

Mite

An animal in the order Acárina, class Arachnida, phylum Arthropoda; including both free-living and parasitic forms.

Mitochondrion

A minute body in the cytoplasm in cells, the chief location of respiratory enzymes. syn. Chondriosome.

Mitosis

The ordinary division of a nucleus that includes the longitudinal doubling of chromosomes to form pairs of chromatids, separation of each pair to form two daughter nuclei; so a constant number of chromosomes is maintained.

Mitotic

Refers to *Mitosis*.

Mixed Forest

A forest composed of trees of two or more species, usually at least 20 per cent of the trees are of other than the leading species.

Mixed Prairie

An extensive grassland type lying west of the tall-grass or *True prairie* in North America, consisting of a mixture of tall-, short-, and midgrasses, and other herbaceous plants.

Mixotrophic

Refers to an organism that is both *Autotrophic* (q. v.) and *Heterotrophic* (q. v.), e. g. *Insectivorous* plants.

Model

The organism that in *Mimicry* is imitated.

Modification

A non-inheritable, *Phenotypic* variation of a characteristic of an organism, caused by the environment, e. g., taller growth of a plant in the shade than in the sun. cf. *Acquired character.*

Moisture Equivalent

The percentage of moisture retained by a small sample of saturated soil after being subjected to a centrifugal force 1000 times that of gravity for a definite period of time, usually one-half hour.

Moisture Stress

The tension at which water is held in the soil.

Moisture Tension

The force at which water is held in the soil.

Mollusk

An organism in the large phylum Mollusca, e. g., snail, oyster.

Monadnock

An erosion remnant such as a hill or a mass of rock rising above the surrounding land.

Monandry

The mating of a female with only one male, or the presence of only one stamen in a flower.

Monoclimax (Theory)

As postulated by F. E. Clements, ecologic succession will

in time culminate in a single *Climax* (q. v.) within a climatic region. cf. *Polyclimax*.

Monoclinous (Monoclinic)

Refers to plants that have perfect flowers, i. e., stamens and pistil in the same flower, e. g., a rose flower. cf. *Diclinous*.

Monocotyledon

A vascular plant in the subclass Monocotyledoneae, class Angiospermae (flowering plants), e. g., grasses, orchids. cf. *Dicotyledon*.

Monoecious

(1) Refers to a plant with some flowers containing only stamens and other flowers with only one or more pistils on the same plant, e. g., corn. cf. *Dioecious*. (2) A unisexual animal or plant. cf. *Hermaphrodite*.

Monogamy

The mating of an animal with only one member of the opposite sex. cf. *Polygamy*.

Monogynous

The mating of a male with only one female. cf. *Polyandry*.

Monohybrid

A cross or *Hybrid* resulting from the mating of parents differing in only one character. cf. *Dihybrid*.

Monolith (Soil)

A sample of a vertical section of a soil profile a few inches thick, removed from the soil with as little disturbance as possible. cf. *Profile (soil)*.

Monophagous

Refers to an organism that subsists on a few or only one

kind of food, e. g. many caterpillars. cf. *Steno-, Euroky.*

Monoploid

See *Haploid.*

Monotopic

Refers to a single area such as in the restricted area of distribution of a species.

Monotreme

An animal in the primitive order Monotremata, egg-laying mammals, restricted to the Australian faunal region, e. g., platypus.

Monsoon

A wind system that reverses its direction with the season, mostly in southeast Asia.

Monsoon Forest

A tropical forest of deciduous trees in regions where seasons of heavy rainfall alternate with long droughts.

Montane

Refers to mountains.

Moor (Moorland)

Primarily high-lying, unenclosed land occupied by heather and other ericaceous dwarf shrubs, including boggy areas. cf. *Bog, Heath, Marsh, Swamp.*

Mor

A layer of *Humus* material, usually compacted or matted or both lying on the mineral soil. cf. *H-layer.*

Moraine

The accumulation of rock material by a glacier, occurs in various topographic forms such as ridges or more level areas according to the manner of formation. Various kinds are lateral, terminal, medial, and ground moraines.

Mores

The general behavioural attributes of motile organisms, or groups of animals possessing particular ecological characteristics.

Morphogenesis

The origin and development of the form and structure of an organism or one of its parts.

Morphology

The study of the form, structure, and development of organisms.

Morphology (Soil)

The constitution of the soil including texture, structure, and other properties.

Mosaic

(1) A pattern of vegetation in which two or more kinds of communities are interspersed in patches, e. g., clumps of shrubs with grassland between. (2) A symptom of some kinds of virus disease.

Moss

(1) A plant in the class Musci, phylum Bryophyta. (2) A *Bog*.

Muck

An organic soil consisting of fairly well decomposed unrecognizable organic material that is finely divided, dark in color, and with a relatively large content of mineral matter.

Mulch

A natural or artificial layer of plant residue or other material such as sand or paper on the soil surface. cf. *Dust mulch*.

Mulch Tillage

Working of the soil so that plant residues are left on the surface.

Mulga

A scrub thicket consisting mostly of *Acacia.*

Mull

A layer of *Humus* that is granular in structure, more or less friable, slightly or not at all matted, and with a gradual transition to the mineral soil below. cf. *Mor.*

Multiparous

Refers to an animal that produces more than one young at birth. cf. *Uniparous.*

Multiple Use

The policy of using a resource in several ways such as the use of forests for the production of timber, forage, water supplies, and game animals, and also for recreation.

Multivoltine

Refers to an organism that has several generations during a single season. cf. *Univoltine.*

Muskeg

A *Bog* in the northern part of North America characterized by an abundance usually of Sphagnum moss and a greater or lesser abundance of shrubs and low trees such as black spruce.

Mutagen

An influence that induces *Mutation* (q. v.) in organisms, e. g., *Ionizing radiation.*

Mutant

An organism, characteristic, or gene resulting from *Mutation.*

Mutation

A sudden, inheritable variation in an organism resulting from changes in a *Gene,* or in alterations of the structure or number of *Chromosomes.*

Mutualism

The kind of interspecies relationship, *Coaction,* or *Symbiosis* that is obligatory and beneficial to the two or more participating organisms, e. g. a fungus and an alga in a lichen.

Mutuality

The relationship where mutual benefit or dependence occurs because of the proximity of organisms to one another.

Mycelium

The mass of *Hyphae* (q. v.) of a fungus, e. g., bread mold.

Mycetophagous

Refers to an organism that eats fungi, e. g., a *Collembolon* that eats *Hyphae.*

Mycetozoa

See *Myxomycete.*

Mycology

The branch of botany that deals with fungi.

Mycorrhiza

The symbiotic relationship of a fungus with the roots of certain plants. cf. *Mutualism, Endotrophic, Ectotrophic.*

Mycorrhizomata

The association of a fungus with a *Rhizome.*

Mycothalli

The association of a fungus with a *Thallus.*

Mycotrophic

Refers to a plant with *Mycorrhiza.*

Myriapod

An animal in the group Myriapoda comprising classes Chilopoda (centipedes) and Diplopoda (millipedes), phylum Arthropoda.

Myrmecodomatia
Structures on plants in which ants or termites live.

Myrmecolous
Refers to an organism that lives in ant or termite galleries.

Myrmecophilous
Refers to plants that are inhabited by ants or termites. cf. *Trophobiosis.*

Myrmecophobous
Refers to plants that repel ants or termites.

Myrmecophyte
A plant that has structures adapted for the shelter of ants or termites and usually also has extrafloral nectaries or glands producing nutritious substances, e. g. *Acacia* spp.

Myxomycete
A slime mold. An organism possessing both animal and plant characteristics, classified in the phylum Myxomycophyta in the Fungi. syn. Mycetozoa.

N

Nanism

The dwarfed appearance of plants, as at tree-line in mountains.

Nanophanerophyte

A subdivision of the *Phanerophytes* (q.v.) in *Raunkiaer's life-form classification,* comprising shrubs 0.25 to 2 meters in height.

Nanophyll

See *Leaf-size classes.*

Nannoplankton

Very minute *Plankton* (q. v.), those that pass through meshes of a No. 20 silk bolting cloth (0.03 to 0.04 mm.).

Nastic Movement

A response in plants caused by a diffuse stimulus (not received from a definite direction) or when the response to a diffuse or lateral stimulus is determined exclusively by the irritable organ, e. g., drooping of the leaves of *Mimosa pudica* when it is touched. cf. *Tropism.*

Natality
The production of offspring by organisms.

Natatorial
Refers to the swimming capacity of an organism.

Natural Area
An area of land in which organisms and geological processes are undisturbed by man, with as few controls as possible. cf. *Primitive area.*

Natural Selection
The agent of evolutionary change by which the organisms possessing certain characteristics in a given environment give rise to more offspring than those lacking such characteristics. cf. *Drift, genetic; Mutation.*

Nature Reserve
See *Natural area.*

Nature Sanctuary
See *Natural area.*

Neap Tides
The lowest tides during a month, occurring about the time of the moon's first and last quarters.

Nearctic
One of the faunal regions of the earth, in the realm *Megagea,* includes North America except the tropical part of Mexico.

Necrosis
The death of an organism or one of its parts.

Nectar
The sweet liquid secreted by special glands in flowers or in other parts of plants, attractive to insects. cf. *Nectary.*

Nectariferous
Refers to a flower or plant that produces nectar.

Nectary

A gland in a flower or on a vegetative organ that produces nectar.

Nekton

The strong-swimming animals in water, e. g., fish. cf. *Benthos, Plankton.*

Nematode

An animal in the class Nematoda, phylum Nemathelminthes, e. g., hookworm, eelworm in potatoes.

Neo-Darwinism

The doctrine of modern evolution that combines *Genetics* with *Natural selection* (q. v.).

Neogea

The faunal realm, containing only one region, the *Neotropical* which includes South and Central America and the tropical parts of Mexico. cf. *Megagea, Notogea.*

Neo-Lamarckism

The theory of evolution that includes modern-day modifications of the doctrines of *Lamarckism* (q. v.).

Neolithic

The cultural stage, beginning about ten thousand years ago, in human history, following the *Paleolithic,* during which cultivation of plants and domestication of animals were started.

Neoplasm

An abnormal increase in the number of cells in some part of an organism, often malignant.

Neoteny

The occurrence of larval or other juvenile characters in the adult stage of an organism, or the presence of an adult character in the larval stage, e. g., larval form of the adult female glowworm. cf. *Paligenesis, Caenogenesis, Paedogenesis.*

Neotropical

See *Neogea.*

Nephometer

An instrument for measuring the percentage of sky that is overcast.

Nephoscope

An instrument for measuring the direction and speed of the movement of clouds.

Neritic

Refers to the portion of the sea lying above the continental shelf, usually to a depth of 200 meters. cf. *Oceanic province, Pelagic, Littoral, Sublittoral.*

Nested Quadrats

An arrangement of placing *Quadrats* in one area so that the size of the area sampled becomes progressively larger, to determine the proper size of quadrat to use for the particular kind of vegetation.

Net-Assimilation Rate

The rate of increase in dry weight of the whole plant in relation to the unit leaf-area or unit leaf-rate.

Neural

Refers to the nervous system or to a nerve.

Neuston

The organisms in a collective sense that are associated with or dependent upon the surface film of water, e. g., mosquito larvae.

Neutralism

The occurrence of two or more populations in an area and neither influences the other.

Névé

Granular, compacted snow at the head of a glacier, or similar snow elsewhere. syn. *Firn.*

Niche

(1) Ecological niche: the role of an organism in the environment, its activities and relationships to the biotic and abiotic environment. (2) Habitat niche: the specific part or smallest unit of a *Habitat* occupied by an organism. cf. *Biotope.*

Nidicolous

Refers to young, undeveloped birds that remain in the nest for a time after hatching.

Nidifugous

Refers to young, undeveloped birds that leave the nest soon after hatching.

Nitrification

The oxidation of ammonia and ammonium compounds to nitrites and then to nitrates by certain bacteria. cf. *Nitrogen cycle.*

Nitrogen Cycle

The circulation of nitrogen, chiefly by means of organisms from the inorganic nitrogen in the atmosphere to nitrates, into proteins and protoplasm in plants and animals, to ammonia, and return to nitrites and nitrates. cf. *Nitrogen fixation, Nitrification.*

Nitrogen Fixation

The assimilation of free nitrogen of the atmosphere by microorganisms in the soil or by bacteria in the nodules of certain plants, especially legumes, into organic nitrogenous compounds.

Nitrophilous

Refers to plants that grow well in soil that is rich in nitrogen, e. g., many barnyard weeds.

Nivation

A kind of erosion caused by the action of snow, e. g. *Solifluction.*

Nocturnal

Refers to night time. cf. *Crepuscular, Diel, Diurnal.*

Nodule

A structure formed on the roots of most legumes and a few other species containing bacteria that carry on *Nitrogen Fixation.*

Nodum

An abstract unit of vegetation such as *Association, Sociation,* and *Alliance;* corresponds to *Taxon* in *Systematics.*

Nomad

A member of a group, especially of primitive people, who change their dwelling place frequently.

Nomina Conservanda

Names of organisms whose usage is maintained by agreement of systematists although the names may be contrary to the rules of nomenclature.

Non-available Water

The amount of water in the soil when a plant wilts permanently. cf. *Wilting.*

Non-reactive Factor

An environmental factor such as weather conditions which are not influenced by the density of individuals in a population, but may produce increasingly adverse effects with increasing density. cf. *Density-independent factor, Density-dependent factor.*

Normal Dispersion, Normal Distribution

The distribution in an area of individuals of a population that is at random.

Normal Erosion

The *Erosion* that occurs on land under natural environmental conditions not disturbed by human activities. cf. *Accelerated erosion.*

Normal Spectrum

See *Biological spectrum.*

Norther (Norte in Central America)

A northerly wind, especially a strong one, that begins suddenly during the colder half of the year in the region from Texas southward, including the Gulf of Mexico and the western Caribbean.

Notogea

One of the three continental faunal realms. It includes the *Australian region* (q. v.). cf. *Megagea, Neogea.*

Nuciferous

Refers to nut-bearing plants.

Nunatak

A body of land, such as a mountain, projecting at some time above a mass of ice and snow, or above a glacier.

Nurse Crop

See *Companion crop.*

Nutation

A circular or spiral movement of the growing portions of plants such as stems or tendrils.

Nutrient (Plant)

Any substance absorbed by a plant that is used in its *Metabolism.*

Nyctinasty

The movement of a plant organ in response to alternation of night and day as in clover leaflets.

Nyctitropism

See *Nyctinasty*.

Nymph

A stage in the *Metamorphosis* of certain insects between the larval and adult forms.

O

Obligate Parasite

A *Parasite* that cannot attain complete development independent of its *Host.*

Oceanic Province

The portion of the ocean seaward from the *Continental shelf,* having a depth greater than 200 meters. cf. *Neritic.*

Oceanography

The science dealing with all aspects of the ocean; physical conditions, plant and animal life, etc.; sometimes restricted to the study of physical conditions only.

Oecology

See *Ecology.*

Oesophagous

See *Esophagous.*

Oestrus (Estrous) cycle

The period in mature females in many kinds of mammals when the desire for mating occurs; it varies in length, is controlled by hormones, and is often accompanied by bodily changes.

Offset

A short, basal stem by which some plants propagate.

Oleaginous

Refers to the production or presence of oil in a plant organ, e. g., olive fruit.

Oligocene

The geological epoch near the middle of the *Tertiary* period in the *Cenozoic* era which began about 39 million years ago and lasted for about 11 million years.

Oligosaprobic

Refers to an aquatic habitat that is high in content of oxygen, low in dissolved organic matter, and with very little decomposition of organic substances that are present. cf. *Catarobic, Polysaprobic, Mesosaprobic.*

Oligotrophic

Refers to ponds or lakes that are low in content of basic nutritive substances for plants, lacking a distinct stratification of dissolved oxygen in summer or winter. cf. *Eutrophic.*

Ombrophilous

Refers to plants that endure much rain, with leaves that are easily wetted.

Ombrophobous

Refers to plants that do not endure much rain, with unwettable leaves.

Omnivorous

Refers to an animal that eats both plant and animal food. cf. *Carnivorous, Herbivorous.*

Ontocline

A gradation in phenotypic characteristics such as color or form appearing at different times in the life cycle of an animal, may be related to the *Ecocline* (q. v.).

Ontogeny

The development of an individual, or a part of it, from the *Zygote* to the adult. cf. *Phylogeny.*

Oogamy

Sexual reproduction by means of eggs and sperms. cf. *Heterogamy.*

Open Community

A community in which the plants are more or less scattered, in which invasion may readily occur.

Open Pollination

Pollination (q. v.) by wind, insects, etc., not directly by man.

Open Range

An extensive grazing area on which the movements of livestock are unrestricted.

Open Woodland

A parkland type of vegetation in which trees do not form an open canopy.

Optimum (Conditions)

The range of conditions which is most favorable for an organism, or for a certain function of an organism. cf. *Ecological amplitude.*

Order

In plant and animal *Taxonomy,* a group or *Taxon* of related families, e. g. Rosales. In phytosociology, a group of similar *Alliances.* In classification of soils, the highest category comprising *Zonal, Intrazonal,* and *Azonal* soils.

Ordovician

A geological period in the early part of the *Paleozoic* era, which began about 425 million years ago and lasted about 65 million years.

Organ
A distinct part of a plant or animal which carries on one or more particular functions, e. g., a leaf, wing of a bird.

Organelle
Any part of a cell of an organism.

Organic Matter (In Soil)
Materials derived from plants or animals, much of it in a more or less advanced stage of decomposition. cf. *Humus.*

Organic Soil
A soil composed mainly of organic matter on a volume basis; containing 20 per cent or more on weight basis, e. g., *Muck, Peat.*

Organismic (Organismal)
Refers to the concept that a group of organisms such as a community has qualities of a higher level of organization than the constituent organisms have individually; changes occurring in a community are related more to the qualities of the group than to those of the individual plants and animals.

Oriental Region
One of the four continental faunal regions in the realm *Megagea;* it includes tropical Asia and the associated continental islands.

Ornithology
The division of zoology that deals with birds.

Ornithophilous
Refers to flowers that are pollinated by birds such as the hummingbird.

Orogenesis
The process of mountain formation by changes in the earth's crust.

Orographic
Refers to mountains, or to relief characteristics of the land.

Ortet
The plant from which members of a *Clone* (q. v.) were derived. cf. *Ramet.*

Orthogenesis
The trend in the evolution of organisms in a particular direction for a long time. cf. *Genetic drift, Sewall Wright effect.*

Orthokinesis
The random movements of an organism that decrease in rate from one part of an environmental gradient to another part, e. g., a meal worm moving from the moist part of a gradient and congregating in the drier part. cf. *Klinokinesis, Klinotaxis, Photokinesis.*

Orthopterous
An insect in the order Orthoptera, e. g., cricket, grasshopper.

Orthoselection
The trend in the evolution of organisms which is in a particular direction under the influence of selection. cf. *Orthogenesis, Natural selection.*

Ortstein
A strongly compacted, indurated layer *(Pan)* of soil in which the particles are cemented together with iron and organic matter.

Osmoregulation
The adjustment in the osmotic concentration of solutes in fluids in organisms to environmental conditions, e. g., when eels migrate from salt to fresh water.

Osmosis

The diffusion of a solvent (especially water) across a differentially permeable membrane separating two solutions or separating a solution and a solvent.

Osmotic Concentration

The concentration of salts in a solution.

Osmotic Pressure (of a Solution)

The rating or index of potential maximum pressure which can develop in a given solution when it is exposed to *Osmosis.*

Osteology

The study of the development and nature of bones in vertebrate animals.

Oued

A valley that contains water in rainy seasons (Arabian).

Outbreak

The occurrence of an organism in large numbers or in sufficient number to cause serious damage over an appreciable area.

Outbreeding

The mating of individuals that are not closely related. cf. *Inbreeding.*

Outcrop

A geological stratum that is exposed on the surface of the earth.

Outwash, Glacial

Material carried by streams of melt water from a glacier and deposited in the form of plains, deltas, and *Valley trains.*

Ovary

(1) The portion of the *Pistil* or *Carpel* of a flower that contains one or more *ovules* (q. v.). (2) The organ in female animals that produces the egg or ovum.

Overdispersion

See *Contagious dispersion.*

Overgrazing

Grazing so intensively that it reduces the capacity of plants to continue production of forage and also causes deterioration by damaging plants or soil or both. cf. *Undergrazing.*

Overpopulation

A population-density in excess of the capacity of the environmental resources to supply the requirements of the individual organisms, usually accompanied by a high mortality rate because of inadequate nutrition, insufficient shelter, and increased predation, disease, or parasitism. cf. *Malthusian.*

Overstocked

(1) Refers to stands of trees in which the large number may retard growth. (2) A population of animals in which the number is in excess of the resources of the habitat to provide food and shelter. cf. *Fully stocked.*

Overstocking

The placing of more livestock on a range than its resources can support through the grazing season without *Overgrazing.*

Overstory

The layer of trees in a forest that forms the *Canopy.* cf. *Understory.*

Overturn

The spring and fall circulation in lakes induced by the

wind when thermally different strata become mixed. cf. *Epilimnion.*

Oviparous

Refers to an animal that lays eggs in which embryos show little or no development, e. g., most fishes. cf. *Viviparous, Ovoviparous.*

Oviposíter

A specialized structure in insects for depositing eggs.

Ovoviparous

Refers to an animal that keeps ova or eggs within the body until they are ready to hatch, requiring internal fertilization, as in birds. cf. *Oviparous, Viviparous.*

Ovule

The structure within the *Ovary* of a flower that after *Fertilization* of the egg within it develops into a seed.

Ovum

A female *Gamete* (q. v.) or egg.

Oxbow Lake

A more or less semicircular lake or pond formed when a meander of a river is separated from the main stream.

Oxidation-reduction Potential

The potential of a given material, in comparison with other materials, to release electrons (oxidation), or to receive electrons (reduction); symbolized by E_h. syn. *Redox potential.*

Oxylophyte

See *Acidophilous.*

Oxyphilous

See *Acidophilous.*

Oxyphobous
See *Basophilous.*

Oxysere
The stages of a successional series that began in water or soil that is appreciably acid. cf. *Succession, Hydrosere.*

P

Paddock

A small *Enclosure* in a grassland.

Paedogenesis

Reproduction occurring in larval or other young stages of an animal. cf. *Neoteny*.

Paedomorphosis

The evolutionary process in which a character of an immature stage of an organism appears in the adult stage. cf. *Neoteny*.

Palatability

The acceptability of food by domestic or wild animals as shown by their preferences; in range management often used for utilization of forage, especially the proper degree of use under good management. cf. *Proper use factor*.

Palearctic

One of the four continental faunal regions in the realm *Megagea* (q. v.). It includes Eurasia north of the Tropics and the northernmost part of Africa.

Paleobotany

The study of fossil plants. cf. *Paleontology, Paleozoology.*

Paleocene

The earliest geological epoch in the *Tertiary* period of the *Cenozoic* era, which began about 75 million years ago and lasted for about 17 million years.

Paleoecology

The study of *Ecology* of former geological periods.

Paleogenic (Palaeogenic)

See *Paleozoic.*

Paleolithic

Refers to the period of human history characterized by food-gathering, fishing, and hunting, without cultivation; and by the use of stone implements.

Paleontology

The study of the life of former geological epochs by means of fossils. cf. *Paleobotany, Paleozoology.*

Paleozoic

One of the major geological eras, preceding the *Mesozoic,* which began about 505 million years ago and ended about 205 million years ago.

Paleozoology

The study of animal fossils. cf. *Paleontology, Paleobotany.*

Palingenesis

The appearance during the development of an organism of stages or structures which occurred in earlier forms during its evolution. cf. *Caenogenesis, Recapitulation.*

Paludel

Refers to marshes.

Palynology

The study of *Pollen* and other microfossils, especially from deposits in lakes and other bodies of water, to determine the age of strata and the kind of plant life existing in former periods. cf. *Paleobotany*.

Pampas

Extensive grasslands in South America, particularly in Argentina, large portions of which are now cultivated. cf. *Prairie, Steppe, Veld.*

Pampero

A suddenly arising. violent, southwesterly wind on the *Pampas* of South America, most prevalent from July to September.

Pan

(1) A layer in the soil that is strongly compacted and indurated, or with a very large clay content. cf. *Ortstein.* (2) A shallow, basin-like depression without vegetation or outlet for drainage.

Panclimax

According to F. E. Clements two or more related climaxes that have the same life-form, common genera of dominants, and the same general climatic factors. cf. *Formation.*

Pandemic

Refers to something which has a wide occurrence such as a disease.

Panmixis

The wide interbreeding of individuals of a population, where each individual has the potential capacity of mating with any other individual.

Pannage

The feeding of swine in woods, or the food such as

acorns that is secured. A term used in England especially.

Pantano

A fresh-water or brackish marsh in Argentina.

Pantograph

A drafting instrument used in copying maps or charts, adapted for charting the location and area of plants in a *Quadrat* (q. v.).

Papagayo

A violent northeasterly wind of the Pacific coast of Central America.

Papilionaceous

Refers to the butterfly-like flower of many species in the pea family, Leguminosae.

Paramo

An alpine type of vegetation, or the sparsely vegetated land, in high mountains in the Andes and in northern South America. cf. *Tundra.*

Parang

A type of mixed vegetation composed of species differing in height, which occurs on areas following the repeated cutting of forest in southeast Asia.

Parasitism

The interaction or *Coaction* in which one or more organisms, the *Parasite,* benefits while feeding upon, securing shelter, or otherwise injuring one or more other organisms, the *Host;* often the term is restricted to nutritive relations. cf. *Symbiosis, Saprophyte.*

Paratonic

Refers to movements of plants such as tropic or nastic movements induced by an external stimulus. cf. *Tropism, Nastic movement.*

Parch Blight

The kind of injury to plants, especially evergreens that have been exposed to strong, drying winds in winter.

Parenchyma

(1) A common tissue in the soft parts of plants consisting of thin-walled, cubical cells, e. g. pith, fleshy fruits. (2) The loose tissue that forms a large part of the body of flat-worms.

Parent Material (Soil)

The *C horizon* (q. v.) of the soil.

Parent Rock (Soil)

The rock from which the parent material of the soil has been formed.

Parkland

The type of landscape in which trees occur in clumps in grassland.

Parthenocarpy

The development of the fruit of a plant without *Fertilization,* e. g., as in the banana.

Parthenogenesis

The development of the egg of an organism into an embryo without *Fertilization.* cf. *Apomixis.*

Pasturage

Herbage of an area taken by an animal when grazing, the yield of a *Pasture.*

Pasture

An area of vegetation used for grazing, sometimes restricted to areas of cultivated land which have been seeded and then used for grazing. cf. *Range.*

Patabiont

An animal that spends its normal life in the debris of the forest floor stratum.

Patana

(1) A montane grassy slope (Ceylon). (2) A grassy slope with a moderate supply of moisture, resembling a *Savanna.*

Pathogen

An organism or virus that causes disease.

Pathology

The study of diseases.

Patocole

An animal that spends a regular part of its normal life outside of the forest floor but lives transiently in it.

Patoxene

An animal that occurs accidentally in the litter of the forest floor.

Pattern

The arrangement formed by the occurrence of individuals or groups of organisms in an area, such as *Contagious dispersion, Hypodispersion,* and *Normal dispersion* (q. v.).

Peat

An organic soil consisting of partially decomposed organic remains which can be identified as to the kind of constituent plants. It accumulates in water or under wet conditions. cf. *Muck.*

Peck-dominance

The dominance of one bird over another in a flock, applied also to other animals. cf. *Social dominance.*

Peck-order

The order of dominance of some birds over others in a flock, applied also to other animals.

Peck-right

The dominance of one bird over all the others in a flock.

Ped

An individual soil aggregate occurring naturally.

Pedalfer

The kind of soil in which alumina and iron oxides move downward in the soil profile and in which no accumulation of calcium carbonate occurs. cf. *Pedocal, Laterite.*

Pedigree

A record of the ancestry of an individual.

Pediment

A gradual slope covered with scattered rock adjacent to a mountain in semi-arid and arid regions.

Pedocal

A type of soil in which a layer of accumulated carbonates occurs. cf. *Pedalfer, Laterite.*

Pedogenic

Refers to effects caused by soil factors.

Pedology

The science dealing with soils.

Pedon

A *Community* dependent on the bottom in lakes. cf. *Benthos.*

Pelagic

Refers to the open water of the ocean, lacking association with the shore or the bottom. cf. *Abyssal, Neritic, Benthic, Littoral.*

Peneplain

A land surface worn down by erosion to almost *Base level* so that most of it appears as a plain.

Penicillin

An *Antibiotic* (q. v.) produced by the mold *Penicillin notatum.*

Pennsylvanian

A geologic period in the latter part of the *Paleozoic* era which began about 255 million years ago and lasted for about 25 million years.

Perched Water

The body of free ground water in a zone of saturation, separated from an underlying body of ground water by an unsaturated layer of material.

Percolation

The downward movement of water in the soil, especially in saturated or nearly saturated soil. cf. *Infiltration*.

Perennate

The continuance of life in an organism from year to year as in a *Rhizome* or seed.

Perennial

A plant that lives for three or more years. cf. *Annual, Biennial*.

Perfect

Refers to flowers containing one or more stamens and pistils.

Perianth

The flower parts surrounding the stamens and petals; the petals and/or sepals of a flower; much reduced in some flowers.

Periodicity

The repeated occurrence of events at fairly frequent and regular intervals. cf. *Aspection, Phenology*.

Periphyton

The assemblage of organisms attached to surfaces submerged in water, above the bottom. cf. *Benthos, Plankton*.

Permafrost

Permanently frozen ground in arctic and subarctic regions.

Permanent Pasture

Grazing land that remains under grazing use for many years. cf. *Rotation Grazing, Range.*

Permanent Quadrat

A marked sample area or *quadrat* (q. v.) in which the vegetation is recorded over a period of many years.

Permanent Wilting Percentage

The quantity of water in the soil on a dry weight basis when plants growing in it have reached the condition of permanent *Wilting.*

Permeability

(1) The property or condition of the soil that relates to the passage of water or air through it. cf. *Percolation.* (2) The rate of diffusion of molecules of a substance through a membrane. cf. *Osmosis.*

Permeant

An animal in a terrestrial community which has great motility, e. g., a bird.

Permian

The latest geological period in the *Paleozoic* era which began about 230 million years ago and lasted for about 25 million years.

Pessimum

The conditions of the environment outside of the *Optimum* which an organism can endure. cf. *Ecological amplitude.*

Pesticide

An agent or substance that destroys pests, e. g., a fungicide or insecticide.

Petal

One of the parts of the *Corolla* of a flower.

Petiole

The stalk of a leaf.

Petran

Refers to the Rocky Mountains.

Petrification

The formation of fossils by the replacement of organic substances in dead organisms by minerals.

pF

A measure of the energy by which a soil holds water; soil dried at 100-110° C has a pF of 7.0, the tension decreases as the water content increases.

pH

A measure of the acidity or alkalinity of a solution, or the concentration of H or OH ions, ranging from 0 to 14. Values above 7 are alkaline, below 7 are acid. cf. *Reaction* (2).

Phage

See *Bacteriophage.*

Phagocyte

A cell that engulfs bacteria or other particles, e. g., white corpuscles in the blood.

Phagocytosis

The engulfing of particles by a *Phagocyte.*

Phanerogam

See *Spermatophytes.*

Phanerophyte

One of *Raunkiaer's life-form classes* (q. v.) in which the buds or other perennating parts are more than 25 cm. above the ground, especially trees and shrubs.

Phenology

The study of the periodic phenomena of animal and plant life and their relations to the weather and climate, e. g. the time of flowering in plants. cf. *Periodicity, Aspection.*

Phenotype

The expression of the characteristics of an organism as determined by the interaction of its genic constitution and the environment. cf. *Genotype, Ecad.*

Phloem

The tissue in plants that conducts foods such as sugar which is performed especially by the sieve tubes.

Phoresy

The transport of one organism by another, e. g., mites carried by insects.

Phosphorescence

See *Bioluminescence.*

Photeolic

See *Nyctinasty.*

Photic Zone

The upper portion of bodies of water into which light penetrates in sufficient intensity to influence plants or animals. cf. *Aphotic zone.*

Photogenic

Refers to the capacity of a substance to produce light. cf. *Bioluminescence.*

Photokinesis

The undirected locomotion of many lower organisms in response to light, accomplished by their capacity to influence the speed of their movements. cf. *Phototaxis, Orthokinesis.*

Photometer

An instrument for measuring the intensity of light.

Photon

One of the particles in a beam of radiant energy. cf. *Quantum.*

Photonasty

The response of an organism or organ to a diffuse light stimulus, e. g., closing of oxalis flowers in the evening.

Photoperiod

The duration of light during a 24-hour period.

Photoperiodism (Photoperiodicity)

The response of plants and animals to the relative duration of light and darkness, e. g. a chrysanthemum blooming under short days and long nights.

Photosynthesis

The synthesis of carbohydrates from carbon dioxide and water by *Chlorophyll* using light as energy with oxygen as a by-product.

Phototaxis

The directed movement of a motile organism in response to a light stimulus. cf. *Photokinesis.*

Phototrophic

Refers to organisms that can obtain energy from sunlight. cf. *Autotrophic, Heterotrophic, Chemotrophic.*

Phototropic

The growing or turning of an organ of a plant when unequally illuminated, toward the light of greater intensity.

Phragmosis

The process of opening or closing holes by animals, e. g., an ant closing its entrance to the nest in a plant stem.

Phreatophyte
A plant that absorbs its water from a permanent supply in the ground, e. g., willows along a stream.

Phycology
See *Algology*.

Phylad
A phylogenetic group of closely related species.

Phylloclade
See *Cladophyll*.

Phylogeny
The evolutionary development and relationships of a group of organisms such as rodents or species of rose. cf. *Ontogeny*.

Phylum
One of the major subdivisions used in classifying plants and animals, e. g., Tracheophyta (vascular plants), Arthropoda (arthropods).

Physiognomic Dominance
Dominance (ecologic) (q. v.) caused by the similarity of a number of species in a certain life-form rather than because of greater cover, number, or size of one or a few species, e. g., a weed patch consisting of many species of annual weeds similar in form.

Physiognomy
The appearance of vegetation as determined by the life-form of the dominant plants, e. g., a grassland, pine forest.

Physiographic Climax
See *Climax*.

Physiography
The branch of physical science that deals with the physical features of the earth. cf. *Geomorphology*.

Physiological Drought

The conditions that obtain when a plant wilts or suffers from insufficient water although the habitat contains ample water. cf. *Aridity, Cold desert.*

Physiological Form

See *Biological race.*

Physiological Isolation

The condition of organisms that have become isolated because of their physiological requirements rather than because of other kinds of *Barriers.* cf. *Isolation.*

Physiology

The branch of biology that deals with the functions and processes carried on by plants and animals.

Physogastry

The condition in some insects in which the body becomes swollen, soft, and white, as in some beetles.

Phytocide

A chemical substance that exerts a differential killing effect within a crop, e. g., one variety but not the other varieties killed by DDT in a barley crop.

Phytocoenosis

The totality of plants in a stand of vegetation, the entire plant *Community.* cf. *Biocoenosis.*

Phytocoenostics

The term preferred by Gams in place of *Plant sociology.*

Phytogeography

See *Plant geography.*

Phytograph

A polygonal diagram that expresses several kinds of community characteristics such as numerical abundance, frequency index, size classes, and basal area on various axes for a certain species of tree in a forest.

Phytometer

A plant such as the sunflower, used to measure environmental conditions.

Phytopathology

The study of plant diseases.

Phytophagous

See *Herbivorous.*

Phytoplankton

The plants occurring in *Plankton* (q. v.), e. g. diatoms.

Phytosociology

See *Plant sociology.*

Phytotron

An extensive set of rooms used for growing plants under controlled environmental conditions.

Pioneer

A plant, animal, or community that first invades a bare area, e. g., willows on a newly formed sandbar.

Pisciculture

The growing of fish.

Pistil

See *Carpel.*

Pistillate

Refers to the pistil, e. g. pistillate flower.

Pitting

The making of shallow pits or depressions, especially in rangeland, with an offset disk or pitting machine, in order to retain rain water or snowmelt.

Pituitary

The gland inside the skull of vertebrates which is of

major importance in the secretion of *Hormones,* many of which control the activity of other *Endocrine* (q. v.) glands.

Plagioclimax

A type of *Climax* (q. v.) vegetation which is the result of man's activity, a biotic climax, e. g., an apparently stable community caused by continued mowing or grazing. cf. *Disclimax, Proclimax, Subclimax, Plagiosere, Serclimax.*

Plagiosere

A *Sere* (q. v.) deflected from its undisturbed course by the constant intervention of man in such activities as burning, grazing, and mowing, resulting in a *Plagioclimax.*

Plain

An extensive tract of nearly level or gently undulating land that is usually occupied by grassland vegetation.

Plane Table

A surveying instrument used for making a sketch-map of a small area.

Planimeter

An instrument used to determine the area of a plane figure or object such as a leaf by tracing the boundaries.

Plankter (Plankt, Planktont)

An individual organism in the Plankton (q. v.).

Plankton

The floating or weakly swimming animal and plant organisms occurring at any depth in lakes, ponds, streams, or seas; often microscopic in size.

Planosol

An *Intrazonal* group of soils with eluviated surface horizons underlain by clay pans or fragipans, developed on nearly flat or gently sloping uplands in humid or subhumid climates. cf. *Eluviation.*

Plant Association
See *Association.*

Plant Community
See *Community.*

Plant Cover
See *Cover.*

Plant Formation
See *Formation.*

Plant Geography
The science that deals with the geographic distribution of plants and the causes of their distribution and dispersal. syn. *Phytogeography, Synchorology.*

Plant Nutrient
The substances or elements absorbed by a plant and used in its metabolism, e. g., nitrates, phosphates. cf. *Essential element.*

Plant Unit
The part of a vegetatively propagating plant which is considered as a unit in analyzing vegetation, e. g., each stalk of plant that has rhizomes.

Plantigrade
An animal that walks on the entire bottom of the feet, e. g., bear, man.

Plant Sociology
The study of plant communities, including their origin, composition, structure, characteristics, distribution, dynamics, and classification. syn. *Phytogeography.*

Plasmagene
Gene-like, non-*Mendelian* carriers of hereditary characters, located in the *Cytoplasm.*

Plasmodium

The mass of cytoplasm containing many nuclei that is enclosed by a single plasma membrane, occurring in *Slime molds* (q. v.). cf. *Syncytium, Coenocyte.*

Plasmolysis

The shrinking of the *Cytoplasm* from the cell wall, caused by osmosis of water out of the cell.

Plasticity

(1) The capacity of an organism to adapt itself to various environmental conditions. (2) The capacity of a soil to be changed in shape under applied stress and to retain the impressed shape after removal of the stress.

Plastid

A protoplasmic body in the *Cytoplasm* in cells of plants, e. g., *Chloroplast, Leucoplast, Chromoplast.*

Playa

Undrained, flat, barren basins that are usually dry and often saline, in arid and semiarid regions.

Pleistocene

The geological epoch preceding the Recent in the *Quaternary* period of the *Cenozoic* era, which began about one million years ago and lasted for about one million years.

Pleomorphic

Refers to an organism that has two or more forms in its life cycle. cf. *Polymorphism.*

Pleuston

A one-layered community of plants that float on or within bodies of water. cf. *Plankton.*

Pliocene

The latest geological epoch in the *Tertiary* period of the

Cenozoic era which began about 12 million years ago and lasted for about 11 million years.

Ploidy

Refers to the number of sets of *Chromosomes* in a cell, e. g., *Diploid, Polyploid* (q. v.).

Plot

An area of land that is studied or used for experimental purposes, in which sample areas are often located.

Plowpan

A compacted layer formed in the soil immediately below the depth of plowing. syn. Plowsole.

Plumule

The portion of the embryo of a seed, above the place of attachment of the cotyledons, consisting of the stem tip and a few embryonic leaves.

Pluvial

Refers to rain.

Pluviilignosa

Vegetation that includes rain forest (Pluviisilvae) which is dominated by evergreen broad-leaved trees, and rain bush (Pluviifruticeta).

Pneumatophore

(1) A special structure on an aquatic or marsh plant that extends above the water making a direct connection of the tissues with the aerial environment. (2) Air sacs exposed to the aerial environment occurring in some aquatic insects.

Pocosin

A bog in shallow, undrained depressions in savannas in the southeastern part of the United States.

Podzol (Podsol)

A *Zonal* group of soils having surface organic layers and

thin, organic mineral horizons above gray, leached horizons upon illuvial, dark brown horizons; developed under coniferous or mixed forests or heath vegetation in cool-temperate, moist climates.

Podzolic

Refers to soils that have part or all of the characteristics of *Podzol.*

Podzolization

The process by which soils are depleted of bases, becoming more acid, and develop leached surface layers from which clay is removed. cf. *Laterization, Calcification, Solonization.*

Poikilotherm

An animal that lacks capacity to control its body temperature which is approximately that of the environment, e. g., a frog.

Point-contact Method (Point-frame)

A technique for determining the area of herbage cover by listing the number of times that various species are touched by the point of a rod or pin. Numerous contacts are used in each *Stand* or *Plot.*

Poisson Distribution

See *Normal dispersion.*

Polar Front

The boundary between the cold air of a polar region and the warmer air of lower latitudes. cf. *Warm front.*

Polarity

The property of an organism to respond differently in its various ends or in contrasting parts of its body to stimuli.

Polder

An area of land reclaimed from the sea or from a lake by dams or dikes, especially in Holland.

Pollen

The *Microspores* which give rise to the male *Gametophytes* in seed plants. cf. *Pollination.*

Pollen Analysis

The identification and determination of abundance of pollen in soil deposits, particularly in peat. cf. *Palynology.*

Pollen Profile

A tabular or graphic expression of the occurrence of various kinds of pollen in deposits of peat or other materials at different depths.

Pollen Spectrum

The expression in percentage of the kinds of pollen in one sample of a *Pollen analysis* (q. v.).

Pollen Tube

The tubular growth, male *Gametophyte* containing the sperms, produced by a pollen grain when it germinates on the stigma and extends into the style of the pistil of a flower.

Pollination

The transportation of pollen by wind, insects, or other agent from an anther to a stigma in flowering plants; usually by wind from a pollen-bearing cone to the *Ovule*-bearing cone in conifers.

Pollution

Contamination of a habitat with substances which make it less favorable for organisms.

Polster

A plant form resembling a cushion, consisting of closely packed stems, dead leaves, and often roots, e. g., many saxifrages, mosses.

Polyandry

The mating of a single female animal with several males. cf. *Monogynous.*

Polyclimax

The concept that several *Climaxes* (q. v.) constitute the vegetation in an area as the result of *Succession* (q. v.).

Polyembryony

In animals the development of more than one embryo from a fertilized egg; in plants the development of more than one embryo within a single *Ovule* of a plant.

Polygamy

The mating of one male animal with several females. cf. *Monogamy, Polygyny.*

Polygonal Soil

The surface configuration of soil characterized by polygons, common in arctic regions.

Polygyny

Mating by *Polyandry* or *Polygamy* (q. v.).

Polymorphism

The presence of several distinct forms in a species, particularly within a certain habitat or population. cf. *Pleomorphic.*

Polyploid

An organism or a cell that contains more than the *Diploid* number of Chromosomes per cell. cf. *Haploid, Tetraploid.*

Polysaprobic

Refers to an aquatic habitat that is characterized by considerable decomposition of organic material and low concentration or absence of free oxygen, cf. *Catarobic, Mesosaprobic, Oligosaprobic.*

Polytopic

Refers to a species that occurs in more than one area, especially to *Discontinuities* (q. v.).

Polytypic

Refers to a species that occurs in various forms in different parts of its range.

Pome

A multicarpellate fruit in which the fleshy part surrounding the ovary is formed from the receptacle, e.g., apple fruit.

Population

A group of interacting individuals of the same species or smaller *Taxa* in a common spatial arrangement.

Population Cycle

The recurrent changes in the size of a population from low to high numbers and the return to low numbers as in the cycle of the snowshoe hare.

Population Density

The number of individuals in a population per unit area.

Population Dynamics

The totality of changes that take place during the life of a population.

Population Pressure

The combined forces of the individuals of a population upon the organisms in a community and upon the environment. cf. *Biotic pressure*.

Pore Space

The fraction of the volume of soil or rock that is not occupied by solid particles.

Porosity

The state of matter which contains *Pore spaces*.

Postadaptation

The increase in *Adaptation* in a *Preadapted* organism after its invasion into a given environment.

Postclimax

A *Climax community* (q. v.) that requires more mesic conditions than obtain generally in the region where it is present, often considered a remnant of a former widespread *Climatic climax* (q. v.).

Potential, Biotic

See *Biotic potential.*

Pothole

A hole formed in rock by the grinding action of a stone kept in motion by a stream.

Potometer

An instrument used to measure the absorption of water by plants.

Prairie

Grassland vegetation, particularly the extensive tract of nearly level or rolling land in North America occupied by tall grasses.

Prairie Soils

A *Zonal* group of soils with dark surface horizons grading through brown to lighter colored parent material at two to five feet, formed under tall grasses in a temperate, humid climate. cf. *Chernozem.*

Preadaptation

The possession by an organism of characteristics that enable it to survive when exposed to new ("prospective") conditions, e.g., the possession of a disease-resistant gene in an organism not yet exposed to the disease. cf. *Postadaptation.*

Preboreal Period

See *Boreal period.*

Precambrian

The geological time preceding the *Cambrian* period, the oldest period of the *Paleozoic* era.

Precipitation

(1) A general term for all forms of falling moisture including rain, snow, hail, sleet, or modifications of them. (2) The quantity of water that is precipitated. (3) The process in which water as a liquid is discharged from the atmosphere upon land or water.

Precipitation-effectiveness Ratio

The total amount of precipitation for a certain period of time divided by the total amount of evaporation, both in liquid form.

Precipitation Rose

A radial diagram expressing the amounts of precipitation by months or other intervals of time.

Preclimax

A *Climax community* (q. v.) that occurs usually in more *Xeric* conditions than obtain generally in the region where it is present, often considered a stage in succession preceding the full development of a climax. cf. *Postclimax.*

Precocial

Refers to young animals that do not need parental care after birth or hatching. cf. *Altricial.*

Predation

The behaviour of animals, *Predators,* in killing other animals, *Prey.*

Predator

An animal that attacks other animals, *Prey,* e.g., a fox that kills mice or other prey.

Predominant

Refers to organisms that are of outstanding abundance or conspicuous importance in a community.

Preferendum

The preferred or "selected" conditions of a motile organism when exposed to a gradient of one or more environmental conditions. cf. *Optimum, Minimum, Pessimum, Tolerance.*

Preferential Species

Species in Class 3 of Braun-Blanquet's classification of *Fidelity* (q. v.); species occurring more or less abundantly, but predominantly or with greater vitality, in a certain community.

Premature Grazing

The grazing of vegetation before the most important forage species have grown sufficiently, or before the soil has become dry and firm enough, to prevent cumulative injury to the range. cf. *Range readiness.*

Prescribed Burning

The use of fire under control to improve growth conditions in vegetation.

Presence

The degree of occurrence of a species in *Stands* of a *Community-type*. cf. *Constancy.*

Pressure, Biotic

See *Biotic pressure.*

Prevernal

Refers to early spring. cf. Aspection.

Prey

An animal that is attacked and killed by another, *Predator* (q. v.), e.g., a ground squirrel killed by the coyote as a predator.

Primary Succession

Succession (q. v.) beginning on a bare area such as a lava flow, not previously occupied by plants or animals. cf. *Secondary succession.*

Primate

An animal in the order Primates which includes lemurs, monkeys, apes, man, and others, in the class Mammalia.

Primeval

Refers to vegetation, geological features, and other natural phenomena, in original condition, before any modification has been made through the influence of modern man, e.g., a primeval forest.

Primitive

Refers to an organism, organ, or behaviour that is characteristic of an early stage in evolution, not specialized or advanced in evolution.

Primitive Area

An area in which conveniences for transportation and living are kept simple and not modernized, cf. *Natural area.*

Primordium

A part of an organism where growth is initiated.

Prisere

See *Primary succession.*

Proclimax

According to F. E. Clements any community that resembles a climax in permanence or extent such as a *Postclimax* and *Preclimax* (q. v.). cf. *Subclimax, Plagioclimax, Disclimax.*

Producer

An organism that can utilize radiant energy to synthesize organic substances from inorganic materials, cf. *Consumer organisms.*

Productivity

(1) The total quantity of organic material produced within a given period by organisms, or the energy that this

represents such as gram-calories per square centimeter per year. (2) The innate capacity of an environment to produce plant and animal life. (3) The capacity of a soil to produce a certain kind of crop under a defined set of management conditions.

Profile (Soil)
A vertical section of soil through all its horizons into the parent material. cf. *A horizon.*

Profundal (Zone)
The body of deep water and the bottom of lakes below the depth of effective penetration of light, cf. *Abyssal.*

Proliferation
The rapid development of new growth such as the production of new parts from buds, offsets, and other organs.

Proliferous
Refers to the ready vegetative reproduction by means of organs such as buds and offsets; the development of leafy shoots from a flower or flower head or other organ.

Propagate
To increase the number of plants vegetatively by bulbs, corms, cuttings, or other plant parts.

Propagule
Any part of a plant that when it is separated from a plant will give rise to a new individual. cf. *Diaspore, Disseminule.*

Proper Stocking
The number of individual livestock in a grazing unit that utilizes the herbage without permanent deterioration of the vegetation or the soil. cf. *Overstocking.*

Proper Use Factor
The maximum percentage of the total amount of annual forage production of a species in a given area within easy

reach of the livestock that may be grazed without permanent deterioration of the plants of this species or associated species nor of the soil. cf. *Overstocking*.

Protandrous

Refers to a flower that produces pollen before the stigma is receptive, e.g., fireweed; or to animals that produce sperms sooner than eggs are produced by the same animal, e.g., certain nematodes. cf. *Protogynous*.

Protective Coloration

The concept that coloration in animals benefits the individual by affording concealment from predators or from prey animals. cf. *Mimicry*.

Protein

A nitrogenous organic compound of large molecular size and complex structure, formed from amino acids.

Proterozoic

The geological era preceding the *Paleozoic* which began about 2,000 million years ago and lasted for about 1,500 million years.

Prothallus

The vegetative structure of the *Gametophyte* (q. v.) generation that is part of the life cycle of ferns and their allies. cf. *Thallus*.

Protista

A group including all one-celled organisms such as one-celled algae, bacteria, and protozoans; suggested as a third kingdom of living organisms, the other two being plants and animals.

Protocooperation

An interaction between organisms that is mutually beneficial but not obligatory to those participating, not applicable to conscious cooperation of human beings. cf. *Commensalism, Symbiosis.*

Protogynous

Refers to a flower that produces pollen when its stigma is no longer receptive to pollen. cf. *Protandrous.*

Protonema

The branched filament of the *Gametophyte* (q. v.) generation that develops from a spore in mosses, and produces leafy branches.

Protoplasm

The living material in cells of animals and plants, usually differentiated into nucleus and cytoplasm.

Protoplast

The organized protoplasmic contents of a cell, used particularly in plants to distinguish the cell wall from the parts within.

Protozoan

An animal, unicellular or non-cellular in the phylum Protozoa.

Provenance

The place of origin of seeds or other *Propagules.*

Psammolittoral

The sandy shore of a lake.

Psammon

The assemblage of organisms that live in the water in the interstices between sand grains in the *Psammolittoral* (q. v.).

Psammophyte

A plant that grows in sandy soil. cf. *Hydrophyte.*

Psammosere

All the stages of a successional series or *Sere* (q. v.) originating in sandy soil. cf. *Hydrosere.*

Psychrometer

An instrument used to measure the *Relative humidity* of the atmosphere by means of the effect of temperature differences of the wet and dry bulb thermometers. cf. *Hygrometer.*

Pteridophyte

A *Vascular plant* in one of the subphyla of the phylum Tracheophyta, exclusive of the seed plants, e.g., fern, club-moss, horsetail.

Public Domain

Land for which the title was originally vested in the Government of the United States by virtue of its sovereignty. More than half of the original public domain has been granted to states, homesteaders, railroads, or has been sold. The remainder is set aside for national forests, national parks, national monuments, Indian reservations, grazing districts, and similar purposes.

Puddled (Soil)

Dense, massive soil artificially compacted when wet, having no regular structure.

Puddling

The process of destroying the structure of the soil during which the porosity and permeability are reduced.

Pulse

The sudden appearance of a great abundance of plant *Plankton.*

Puna

The cold, bleak portions of the high plateau region in the central part of the Andes in South America.

Pupa

The stage in the *Metamorphosis* of an insect between the larval and adult stages.

Pupation

The process during the *Metamorphosis* of an animal when the *Pupa* is formed.

Pure Line

A series of generations of individuals which orginated in a *Homozygous* ancestor.

Puszta

A grassland type of vegetation in the Plains of Hungary. cf. *Steppe, Prairie.*

Pyramid of Numbers

The concept of C. Elton that in most *Food-chains* the number of individuals decreases in each succeeding stage, large numbers of animals occur at the base, a few large ones at the top.

Pyrheliometer

An instrument for measuring solar radiation.

Pyrrhic

Refers to fire.

Q

Q_{10} Rule, Van't Hoff Rule

The rate of response of a process in an organism is often doubled or more for each increase of 10°C. of temperature within certain limits.

Quadrat

A sampling area, originally square, most commonly one square meter, used for analyzing vegetation. A major quadrat is usually 10 meters square. cf. *Plot, Permanent quadrat.*

Quadruped

An animal with four feet, e.g., cow.

Quagmire

Soft, wet, boggy ground which quakes or yields underfoot. cf. *Bog.*

Quantum

A unit of energy that is emitted by a *Photon* (q. v.).

Quaternary

The latest geologic period of the *Cenozoic* era which began about one million years ago, includes the *Recent* and *Pleistocene* epochs.

Quotient of Similarity

See *Index of similarity.*

R

Race

(1) Biology. A *Population* within a species that differs in one or more inherited characteristics from other populations but not sufficiently distinct to rate as a *Taxon* (q. v.). (2) A rapid movement of the tide through a narrow channel.

Rachion

The line where waves break in lakes, the places where wave action and undertow cause the greatest turmoil.

Rad

The dose or unit of ionizing radiation absorbed by tissues of an organism, equal to 100 ergs of energy per gram.

Radiation

(1) The emission and transmission of energy from a source, e.g., *Ionizing radiation* (q. v.), solar radiation by electromagnetic waves such as light, x-rays, gamma-rays. (2) See *Adaptive radiation.*

Radicle

The lower end of the axis of the embryo in a seed and which develops into a root during germination.

Radioactivity

The spontaneous breakdown of certain atomic nuclei usually resulting in the emission of radiant energy in the form of *Alpha* or *Beta particles,* or *Gamma rays.*

Radioautograph

A representation of an object containing radioactive *Isotopes* (q. v.) such as a leaf on a photographic negative to show the distribution of the radioactive material.

Radiocarbon Dating

Determination of the age of organic remains such as long-buried wood by measuring its radioactivity caused by C^{14}, which has a half-life of 5568 years and begins to break down upon the death of organisms.

Radioisotope

An *Isotope* (q. v.) that is unstable, disintegrates, and emits radiations, e.g., uranium-235 which emits alpha and gamma rays.

Radiometer

An instrument that measures the intensity of solar *Radiation.* cf. *Pyrheliometer.*

Radiosensitivity

The sensitivity, or lack of tolerance of organisms to endure *Ionizing radiation* (q. v.).

Radiosonde

A free balloon with attached instruments and radio transmitter for securing measurement of temperature and atmospheric pressure.

Rainfall

The total amount of *Precipitation* including rain, snow, hail, and other forms.

Rain Forest

A type of vegetation consisting of tall, evergreen trees,

mostly broadleaved, occurring in equatorial regions with much rainfall and no, or very short dry seasons.

Rain Gage

An instrument to measure the amount of rainfall.

Rain Shadow

Refers to an area in which little or no rain falls because it is located to the leeward of mountains which on the opposite side are exposed to moisture-laden winds.

Raised Beach

The shore of a former lake or sea that has been elevated by a movement of the earth to form a narrow plain.

Raised Bog

A *Bog* with vegetation of *Sphagnum* spp. and associated plants that is typically convex and gently sloping from the center toward the steep margins, and bordered by a ditch or a watercourse (lagg).

Ramet

An individual member of a *Clone* (q. v.). cf. *Ortet.*

Random Sample

A sample of plant or animal life, soil, or other material or objects in an area in which the sample is located spatially by chance or at random; in contrast to selected samples or *Systematic sampling* (q. v.).

Random Searching

The hypothesis that populations obtain food, suitable niches, and mates by entirely unorganized search, in contrast to systematic searching.

Range

(1) The extent of the geographic area in which a plant or animal occurs. (2) Land covered with plants that are suitable for grazing; usually extensive in area and not suitable for cultivation, especially in arid, semiarid, or forested

regions. cf. *Pasture*. (3) A unit of grazing land used by a given herd of livestock.

Range Condition

The status of vegetation and soil of a given range area in relation to the optimum status (considered by some the *Climax*) obtainable under the prevailing environmental conditions.

Range, Home

See Home range.

Range Improvement

Any procedures that are used to increase the value and ease of management of the range such as the development of water supplies, fencing, revegetation, and control of undesirable plants.

Range Management

The handling of range land to obtain the continuous production of forage and livestock, consistent with uses of the land for other important purposes. cf. *Proper stocking*.

Range Readiness

The degree of growth of important forage plants on the range and the condition of the soil so that livestock may graze without undue compacting of the soil or decreasing the capability of the plants to maintain themselves. cf. *Premature grazing*.

Range Survey

A systematic and comprehensive analysis and inventory of the resources of a range area and the related problems of management, and the formation of plans for management of these resources.

Raptorial

Refers to a predatory animal that has feet with curved, sharp claws adapted for seizing *Prey*, e.g., eagles.

Rasorial

Refers to an animal that usually scratches the ground for food, e.g., barnyard fowl.

Rassenkreis

A series of races in which distinct forms of a *Polytypic* species replace each other in a geographic progression. cf. *Race* (1).

Ratoon

A shoot from a perennial plant such as sugarcane.

Raunkiaer's Law of Frequency

See *Frequency, law of.*

Raunkiaer's Life-form Classification .

A system of classification of *Life-forms* of plants based on the kinds and position of the organ with respect to the soil level that survives unfavorable environmental periods. cf. *Phanerophyte, Geophyte.*

Ravine

An elongated, narrow depression, larger than a gully, usually formed by running water.

Raw Humus

See *Mor.*

Reaction

(1) The effects which one or more organisms produces upon its habitat. cf. *Interaction.* (2) The degree of acidity or alkalinity of a substance, usually expressed as *pH* (q. v.). cf. *Hydrogen-ion concentration.*

Reaction Time

The time required by an organism for the manifestation of a response to a stimulus.

Realm

One of the major divisions in the classification of continental *Faunas* according to P. J. Darlington, Jr., e.g., *Megagea* (Arctogea), *Neogea, Notogea* (q. v.).

Recapitulation

The theory that in the development of an individual the stages of earlier forms in its evolution are repeated, e.g., gill slits in the embryo of a pig. cf. *Palingenesis.*

Recessive

Refers to a *Gene* that has no effect on the appearance of an organism unless it is Homozygous (q. v.), i.e., the dominant gene is not present. cf. *Allele.*

Redox-potential

See *Oxidation-reduction potential.*

Red Tide

See *Dinoflagellate.*

Reduction Division

See *Meiosis.*

Reed Swamp

A community of plants such as cattail or bulrush that grows in shallow water and often extends landward on wet soil where the water table is near the surface.

Reef

A series of rocks close to the surface of a body of water, may be exposed at low tide.

Reflex

An innate, simple, stereotyped response, located in the nervous system, occurring very shortly after the stimulus has been received by an animal.

Reforestation

The establishment of a forest on previously cleared land. cf. *Afforestation.*

Refuge

An area designated for the maintenance of animals within which hunting or fishing is prohibited or strictly controlled. cf. *Natural area.*

Refugium

An area that has not been exposed to great changes undergone by the region as a whole, and as a result provided conditions suitable for the survival of *Relic* (q. v.) species. cf. *Nunatak.*

Reg

Parts of the Sahara desert with a gravelly or stony surface because the fine materials have been blown away.

Regeneration

The process occurring in some animals by which a part of the body which has been lost may be restored, e.g., a crayfish growing a new appendage, growth of new tissue in the wound of a mammal.

Regional Climax

See *Climatic Climax.*

Regionalism

The concept of the integration of relations in human society to the particular conditions and resources that obtain in a region such as the Columbia River basin or the northern Great Plains.

Regolith

The unconsolidated mantle of soil material and weathered rock on the surface of the earth.

Regosol

A deep soil consisting of loose material without stones and shows only slight development of a *Profile* (q. v.).

Regression

(1) A statistical method for the study and expression of the change in one variable associated with and dependent upon changes in another related variable or group of variables. (2) See *Retrogression.*

Regulatory Mechanism

Any influence in the physical or biotic environment of a population that tends to maintain the number of individuals which the resources of the habitat can support. cf. *Reaction, Coaction, Predation, Parasitism.*

Relative Humidity

See *Humidity, relative.*

Relevé

An analysis sample of a *Stand* of vegetation in which are given the data on characteristics such as kinds of species, *Cover, Density,* and sometimes others. cf. *Sample area.*

Relic (Relict)

(1) A remnant or fragment of a flora or fauna that remains from a former period when it was more completely developed. (2) A remnant of the population of a species that was formerly more widespread.

Relief

The inequalities in the elevation of the land surface. cf. *Topography.*

Rendzina

An *Intrazonal* group of soils usually with brown or black, friable surface horizons and light gray or pale yellow calcareous material below, formed under grassland or mixed grassland-forest in humid or semiarid regions. cf. *Prairie soil.*

Replacement Control

The substitution of one kind of plant by another kind, chiefly as a result of competition, e.g., downy bromegrass which is resistant to leafhoppers replacing Russian thistle and mustards which are susceptible to the insect. cf. *Succession.*

Reproduction, Vegetative

Propagation of a plant by stems, roots, or other asexual organs, e.g., strawberry plant increasing in number by means of runners. cf. *Propagate, Sexual reproduction, Asexual reproduction, Apomixis.*

Reproductive Isolation

The separation of populations or organisms so that interbreeding cannot occur.

Reproductive Potential

The maximum rate of increase in numbers of individuals of a species or a population under the most optimum conditions, in contrast to the actual reproduction obtained under existing conditions. cf. *Biotic potential, Environmental resistance.*

Reptile

An animal in the class Reptilia of the subphylum Vertebrata, e.g., snakes, crocodiles.

Réseau

The group of meteorological stations operating under a common direction or in the same territory.

Residual Soil

A kind of soil formed in place by the disintegration and decomposition of rocks and the consequent weathering of the mineral materials. cf. *Alluvium.*

Residual Stand

See *Second growth.*

Resistance

The capacity of an organism to remain relatively unaffected by insects, disease-causing bacteria or fungi, or severe conditions in the physical environment because of inherent qualities that it possesses.

Respiration

The complex series of chemical reactions in all living organisms by which the energy in foods is made available for use. In aerobic respiration free oxygen is utilized and carbon dioxide is liberated; in anaerobic respiration, free oxygen is not required.

Respiratory Quotient

The ratio of the volume of carbon dioxide given off by an organism to the volume of oxygen used in a unit of time.

Restocking

The state in a forest in which replacement of important trees is taking place by natural or artificial means. cf. *Reforestation.*

Retrogression

The change from a more highly organized individual, group, or state of organization to one on a lower level, as in a *Succession* (q. v.) that recedes from the *Climax.*

Revetment

A structure or obstacles placed along the margins of a stream in order to protect the bank from erosion.

Rheology

The division of *Limnology* (q. v). dealing with running waters, their physical, chemical, and biological conditions and interrelationships. cf. *Lotic.*

Rheophyte

A plant that grows in running water. cf. *Hydrophyte.*

Rheotaxix

The orientation in the locomotion of organisms in streams with reference to the current.

Rheotropism

The response of an organism to a current.

Rhesus Factor

See RH factor.

RH Factor

An *Antigen* (q. v.) found in the blood of a large number of human beings who consequently possess *Antibodies* against the antigen. Syn. Rhesus factor.

Rhizine

An organ that serves for the attachment of certain organisms such as the many intergrown filaments in many lichens.

Rhizobia

Bacteria that occur in the nodules of certain plants, especially in the pea family, Leguminosae, and fix free nitrogen into forms useful to the *Host*.

Rhizoid

A filamentous organ, one cell thick, found in mosses, fern *Gametophytes,* and other plants, used for attachment and probably also for absorption of water and nutrient salts.

Rhizome

An underground stem that produces shoots and roots at the nodes. cf. *Rootstock, Runner.*

Rhizosphere

The portion of the soil close to and under the influence of the root of a plant.

Rhythm

The more or less regular recurrence of phenomena such as day and night, differences in animal behaviour. cf. *Periodicity.*

Ridge Terrace

A long, low ridge with gently sloping sides and a shallow channel along the upper side for the purpose of collecting run-off water and diverting the flow across the slope, thus controlling erosion. cf. *Bench terrace.*

Rift Valley

A long, narrow valley between two approximately parallel geological faults, e.g., the extensive one extending from Lake Nyasa northward in the eastern part of Africa.

Rill Erosion

The removal of soil by running water resulting in the formation of shallow channels that can be smoothed completely by cultivation in the normal manner. cf. *Gully erosion, Sheet erosion.*

Rime

A feathery or rough layer of ice deposited on plants and other objects by a fog. cf. *Hoarfrost.*

Riparian

Refers to land bordering a stream, lake, or tidewater.

Riprap

Stones placed on the face of a dam or on stream banks or other land surfaces in order to protect them from erosion; often applied also to other materials that are used for erosion control.

Riverwash

Barren alluvial land exposed along streams at low water levels.

RNA (Ribose Nucleic Acid, Ribonucleic Acid)

A nucleic acid occurring in the cytoplasm of plant and animal cells.

Roaring Forties

Strong westerly winds over the ocean between latitudes 40°S. and 50°S., or the region in which these winds occur.

Rock Flour

Fine material formed by the action of a glacier grinding rocks in its base as it moves forward.

Rodent

An animal in the order Rodentia in the class Mammalia, e.g., mouse.

Roentgen

(1) A unit of measurement of radiation caused by *Gamma rays* and *X-rays,* very similar to *Rad* (q. v.). (2) Refers to *X-rays.*

Rogue

To remove undesired individuals from a population to prevent their reproduction.

Rookery

The breeding place of a group of birds or seals.

Root Climber

A plant that ascends by means of roots attached to a support, e.g., poison ivy on the trunk of a tree.

Root Nodule

See *Nodule.*

Rootstock

See *Rhizome.*

Root Sucker

A sprout arising from a root.

Root Zone

The part of the soil occupied by roots, or subject to such occupation under normal conditions. cf. *Rhizosphere.*

Rotation Grazing

The orderly alternation in the use of two or more portions of a range or pasture.

Rotifer

An aquatic animal in the phylum Rotifera, possesses circles of cilia at the anterior end.

Rouches Moutonnées

Mounds of smoothed rock marked by striations caused by a glacier passing over them.

Roughage

Feed for animals consisting of plants or plant parts contains a high fiber content and low total digestible nutrients, e.g., hay, *Stover.*

Rough Grazings

Grasslands, largely on hills and mountains in Great Britain, which have replaced forest by natural processes, maintained as grassland by grazing, in contrast to pastures seeded by man.

Roundworm

See *Nematode.*

Row Crop

A crop such as corn planted in rows relatively far apart, usually two to four feet, to permit cultivation between the rows.

Ruderal

A plant inhabiting fields or waste places. cf. *Weed, Exotic.*

Rudimentary

Refers to an organism or part of one that is in an early stage of development or evolution.

Rumen

The first stomach of a *Ruminant* (q. v.).

Ruminant

An animal in the order Artiodactyla, even-toed ungulates,

class Mammalia, that lacks upper incisor teeth and chews the cud, e.g., giraffe, ox, deer.

Runner

An above-ground, more or less horizontal stem that forms roots and shoots at some of the nodes under favorable conditions, e.g., the strawberry plant, Bermudagrass. cf. *Rhizome, Stolon.*

Runoff

The part of *Precipitation* which as surface run-off flows off the land without sinking into the soil and the part that enters the ground and passes through into surface streams as groundwater run-off.

S

Sabalian Life Zone

The portion of the *Austral life zone* (q. v.) which borders the Gulf of Mexico from the southern tip of Florida to the 17th meridian.

Sabulicole

A sand-dwelling insect.

Salina

A salt flat.

Saline Soil

A soil that contains soluble salts, usually chlorides and sulfates in high enough concentration so that the growth of most crop plants is reduced, pH is less than 8.5, often called "white alkali" because of the presence of a white or gray crust on the surface. cf. *Alkali soil, Solonchalk.*

Salinity

The quality of saltness in seawater or fresh water, most commonly expressed in parts of dissolved salt per 1000 parts of water, e.g., salinity of seawater is 35 parts per thousand. cf. *Alkalinity*.

Salinization

The formation of a saline soil by the addition of salts to a non-saline soil, as occurs often in irrigating land with water of a high degree of *Salinity*.

Saltation

A sudden change.

Saltatorial (Saltatory)

Refers to leaping or dancing, e.g., the hind limbs of a kangaroo adapt it to saltatorial motion.

Salt Grounds

Places in a pasture or on a range where salt is placed for consumption by livestock.

Saltigrade

Refers to an animal possessing legs that are adapted for leaping, e.g. kangaroo rat.

Salt Marsh

A *Marsh* in which the water is salty or brackish, with greater *Salinity* than fresh water.

Salt Pan

A depression in a salt marsh, usually bare of vegetation.

Sample Area

A portion of an area of vegetation or of a plot that is used for sampling purposes. cf. *Plot, Quadrat.*

Sample Plot

A measured area in vegetation used for sampling or an area of land used for experimentation.

Sample, Random

A sample taken without bias from an area or from a population in which every part of the area or population has an equal chance of being taken.

Sample, Representative

A sample taken that is typical of or that represents a fair value of the area or population from which it is taken.

Sand Binder

A plant that holds sand from being blown away.

Sapling

A tree that is more than three feet in height and less than 4 inches in diameter at breast height.

Saprobe

An organism that lives on dead organisms or on decaying organic material.

Saprophyte

A plant that obtains food from dead or decaying organic material. cf. *Parasite, Heterophyte.*

Sapwood

The outermost part of the wood or xylem of the trunks of trees, generally lighter in color than the heartwood, contains living cells, active in translocation of water and mineral nutrients.

Saturation Deficit

The difference between the pressure of water vapor in the atmosphere at a given time and the maximum that it could contain at the same temperature, expressed in millimeters of mercury; or sometimes expressed as the difference between the relative humidity and the humidity at saturation.

Saturation Point

The density of a given population above which it no longer increases.

Saurian

Refers to a lizard.

Savanna (Savannah)

An area of grassland in which are scattered trees or shrubs but little or no breaks in the continuity of grassland cover.

Saxicolous

Refers to organisms growing on or among rocks, e.g., many mosses and saxifrages.

Scandent

Climbing.

Scansorial

Refers to an organism that has adaptations for climbing, e.g., a *Liana* such as Virginia creeper. cf. *Ambulatorial, Fossorial.*

Scarp

See *Escarpment.*

Scatology

The study of *Scats,* identification, determination of contents, etc.

Scats

Animal feces or droppings such as pellets of rabbit dung.

Scavenger

An animal that eats animal wastes and dead bodies of animals not killed by itself, e.g., vulture.

Schizomycete

A plant in the phylum Schizomycophyta, bacteria.

Scion

Any aerial plant part, often a small branch, that is grafted onto the root-bearing part (*Stock*) of another plant.

Sciophyllous

See *Heliophyllous.*

Sclerenchyma
Thick walled cells, fibers or stone cells, constituting strengthening tissue in plants.

Sclerophyll
Plants with stiff, leathery, evergreen leaves, may be broad-leaved as the holly, or narrow-leaved as the pine. cf. *Chaparral.*

Sclerosis
The hardening of tissue by an increase in the content of lignin in plants, or an increase in collagen in animals.

Sclerotium
A hard, compact, rounded mass of hyphae serving as a dormant stage for carrying a fungus through an unfavorable environmental period; it may survive many winters in the soil.

Scree
See *Talus.*

Scrub
Densely growing, low, often stunted bushes or trees, cf. *Fruticeta, Bush, Chaparral.*

Searching
See *Random searching.*

Season, Critical
The part of the year when a species suffers greatest mortality, e.g., the migration time of some migratory birds.

Seasonal Aspects
See *Aspection.*

Seaweed
An alga, usually large, growing in the sea, e.g., kelps.

Sebaceous
Refers to fatty material, particularly to the gland in the skin of mammals that secretes sebum.

Secondary Sexual Character

A characteristic, not directly associated with the characters directly connected with reproduction, in which the male and female animals differ, such as the difference in coloration of many male and female birds.

Secondary Species

The species in a community that are subordinate to the *Dominant species* (q. v.), e.g., dogwood shrubs or trees in a white oak forest.

Secondary Succession

The kind of Succession (q. v.) which takes place following the destruction of part or all of the vegetation in an area, usually caused directly or indirectly by man. cf. *Primary succession, Sere, Plagiosere, Subsere.*

Second-foot

A measuring unit for the volume of the flow of water expressed in cubic feet per second.

Second Growth

(1) A forest which comes up after the removal of the old stand by cutting, fire, or other cause; actually *Young growth.* (2) The smaller trees left after cutting all the merchantable trees, actually *Residual stand.* cf. *Virgin forest, Advance growth.*

Secretion

The process of passing substances made within a cell, particularly gland cells, to the outside of the cell; or the substance itself, e.g., nectar from nectar glands in flowers.

Sedentary

Refers to an animal that has little tendency to move about, e.g., certain spiders. cf. *Sessile.*

Sedge

A plant, resembling a grass in vegetative appearance, in

the family Cyperaceae, with usually solid stems, three-ranked leaves, and closed leaf sheaths.

Sedimentary Rock

Rock formed from materials such as pebbles, sand, and clay in rivers, lakes, and seas; usually in distinct layers, e.g., conglomerate, sandstone, limestone.

Sedimentation

The process of depositing materials from a liquid, especially in bodies of water. cf. *Sedimentary rock, Alluvium.*

Sediment Delivery Ratio (Percentage)

The relation of the annual *Sediment yield* to the annual gross amount of erosion.

Sediment Yield

The total sediment outflow from a watershed, includes coarse and fine materials, bedload, and suspended materials; a part of the gross erosion from an area.

Seed

The reproductive structure in Spermatophytes (flowering plants and conifers) formed from the *Ovule* (q. v.), containing an *Embryo,* seed coat, and, in many kinds of plants, an *Endosperm.*

Seed Leaf

See Cotyledon.

Seedling

A young plant produced from a seed. Usage in forest nurseries; a tree which is still growing in the nursery, not transplanted. Usage in forest reproduction; a tree grown from seed and less than three feet in height.

Seed Plant

A *Spermatophyte* (q. v.).

Seed Tree

A tree purposely left standing at the time of cutting a forest, for the purpose of producing seed for reproduction of trees in the surrounding area.

Seed Year

A year in which a given species bears seed in large numbers.

Seepage

(1) The water that passes through or emerges from the ground along a line or surface in contrast to a spring where the water emerges from a localized spot. (2) The process by which water passes through the soil.

Segregation

Separation of the *Gene* pairs (*Alleles*) and distribution of each gene into separate cells during *Meiosis* (q. v.).

Seiche

An oscillation in the level of the surface of a lake or inland sea.

Selection

See *Artificial selection, Natural selection.*

Selection Pressure

A criterion of the results of *Natural selection* (q. v.) upon a population.

Selective Cutting (Felling)

The system of removing certain trees such as the largest ones in a forest. cf. *Clear cutting.*

Selective Grazing

The habit of grazing animals to eat certain plants in preference to others. cf. *Palatability.*

Selective Species

The group of *Characteristic species* (q. v.) in Braun-

Blanquet's scheme of classification that contains species occurring most frequently in only one kind of community, but may rarely occur in others also.

Seleniferous

Refers to soils, plants, and other substances that contain a relatively high concentration of selenium. Some species of *Astragalus* are toxic to livestock because of accumulations of selenium in the tissues.

Self-fertile

Refers to an organism in which *Self-fertilization* occurs.

Self-fertilization

The process by which eggs of an organism can be fertilized by sperms produced by itself. cf. *Hermaphrodite, Self-sterility.*

Self-incompatibility

The condition where pollen of a plant fails to develop effectively on a *Stigma* of the same plant. cf. *Compatibility, Self-sterility.*

Selfing

A process of causing *Self-pollination.*

Self-pollination

The transfer of pollen from an *Anther* to the *Stigma* in flowers on the same plant. cf. *Cross-pollination.*

Self-pruning

The death and falling of branches, especially the lower ones, of living trees.

Self-sterility

The failure of an egg of an organism to be fertilized by a sperm produced by the same organism. cf. *Self-incompatibility.*

Selva

An equatorial *Rain forest* in the Amazon River basin in South America.

Sematic

Refers to an organ, odor, color, or other attribute of an animal that may serve as a warning to other animals.

Semiarid

Refers to a region or climate that is intermediate between *Arid* and *Subhumid,* with *Precipitation-effectiveness ratios* ranging between 16 and 32, and supporting grassland or shrub types of vegetation.

Seminal

Refers to seed or semen.

Seminatural Community (Vegetation)

A community in which the development or characteristics have been modified in part by man's influence, e.g., a successional community on an area where sagebrush was burned by man.

Semipermeable

Refers to a membrane that permits certain substances to pass through readily while others pass through slowly or not at all; differentially permeable.

Senescence

The process of aging.

Senility

The state of old age.

Sensory

Refers to the capacity or sense-organ of an animal by which it receives stimuli.

Sepal

See *Calyx*.

Seral

Refers to *Sere* (q. v.).

Serclimax

A stage in a *Sere* before the *Climax* is reached and remains in that stage indefinitely, e.g., tule marshes in California, cf. *Subclimax, Plagioclimax.*

Sere

The series of stages that follow one another in an ecologic *Succession* (q. v.). cf. *Hydrosere, Xerosere, Subsere, Primary succession.*

Serology

The study of the reactions of blood serums to the introduction of foreign substances into the body of organisms.

Serotinal

Refers to the latter part of the summer season. cf. *Aspection, Estival.*

Serotinous

Refers to late opening such as cones of some pine trees which remain on the trees for several years without opening.

Serpentine

A rock or mineral consisting of hydrated magnesium silicate.

Sessile

An organism that is attached to an object or is fixed in place, e.g., barnacles. cf. *Sedentary.*

Seston

The aggregate of substances and organisms that float or swim in water, including the *Bioseston* (living organisms) and the *Abioseston* (non-living).

Settling Basin

The widening or deepening of a stream so that materials carried in suspension will be deposited.

Sewall Wright Effect

The postulate that if a population is subject to cyclical fluctuations in abundance, the evolutionary trend of the species is influenced by the size (population density and area inhabited) of the minimum breeding population. cf. *Natural selection.*

Sex Chromosome

A *Chromosome* (q. v.) that determines the sex of the offspring of an organism.

Sex-linkage

The connection or association of certain *Genes* or attributes of an organism with the *Sex chromosome,* e.g., certain kinds of color blindness.

Sex Ratio

The relationship of the numbers of males and females in a population, approximately 1:1 in most kinds of animals.

Sexual Dimorphism

The condition in which marked differences in characteristics such as color, size, and form occur between male and female animals in the same species.

Sexual Reproduction

The production of offspring resulting from the fusion of sex cells (*Gametes,* eggs and sperms). cf. *Asexual reproduction, Propagate.*

Sexual Selection

A theory to explain certain kinds of evolution based upon selective mating such as the choosing of a certain mate by a female animal because of attractive features such as the bright coloration of the male.

Shade Plant

A plant that can grow in the shade. cf. *Heliophyllous.*

Shale

An easily splitting sedimentary rock formed from clay or silt.

Shamal

A wind that blows with considerable constancy, carrying much dust, in Iraq.

Sheep Month

The amount of forage or feed required for maintenance by a mature sheep or an ewe and its suckling lamb for 30 days, usually figured as equivalent to one-fifth of a *Cow month* (q. v.).

Sheet Erosion

The erosion or removal of a rather uniform layer of soil from the surface of the land by rainfall and *Run-off* water. cf. *Gully erosion, Rill erosion.*

Shelter

The conditions, objects, or material that provide organisms suitable resting places or protection from attack by predators or from unfavorable conditions of the physical environment. cf. *Covert.*

Shelterbelt

A long *Windbreak* of living trees and shrubs extending over an area larger than a single farm.

Shingle

More or less rounded pebbles varying in size, often mixed with sand, on seashores.

Shinnery
Vegetation consisting of dwarf oaks as dominants, especially in sandy areas in the southern Great Plains.

Shipworm
An animal of the Lamellibranchiata (clams), especially *Teredo navalis* which burrows in submerged wood.

Shock Disease
The deterioration in overcrowded populations of an organism in which the activity of the *Endocrine glands* (q. v.) is abnormal and the general condition and viability of the animals are reduced, which may result in a rapid decline (crash) in the number of individuals.

Shoot
The stem and leaves of a plant taken collectively. cf. *Sprout*.

Shore, Depositing
The addition of sand, silt, and clay, or the removal of rock, to the land adjacent to a body of water by the action of water or wind.

Short-day Plant
A plant that blooms when periods of light are short and periods of darkness are long, e.g., chrysanthemum. cf. *Photoperiodism*.

Short Grass
Grasses that grow only a few inches high, particularly blue gramagrass and buffalograss. cf. *High grass, Medium-height grass.*

Shrub
A perennial woody plant that differs from a tree by its low growth and the possession of several stems arising from the base.

Siblings (Sibs)

The offspring, brothers and sisters taken collectively, from the same parents.

Sibljack

Shrub vegetation on deforested land in the Balkan Peninsula.

Siccation

Processes that include the diminution of rainfall (desiccation) and the drying out of the earth's crust and atmosphere. (*Exsiccation,* q. v.)

Siccideserta

Dry areas such as steppe and desert occupied by open vegetation.

Siccocolous

See *Xerophilous.*

Sierozem

A *Zonal* group of soils with brownish gray surface horizons that grade through lighter colored material to a layer with accumulated calcium carbonate, in arid temperate climates where vegetation usually is shrubby.

Sierra

A chain of mountains with jagged tops.

Sieve Tube

A tube of cells connected end to end, part of the *Phloem* tissue in plants, used to conduct food.

Silage

The partly fermented above-ground parts of crops such as corn, sorghum, legumes, or grasses preserved in a succulent condition for feeding livestock.

Silt

Mineral particles in the soil, intermediate between clay

and sand; 0.5 to 0.002 mm. in diameter according to the U. S. Department of Agriculture system, 0.02 to 0.002 mm. in diameter according to the International system. (2) In a general sense waterborne sediment in which the diameters of individual grains are similar to those of silt (1). (3) Soil material containing 80 per cent or more of silt (1) and less than 12 per cent of clay.

Silting (Siltation)

The deposition of water-borne sediments in bodies of water, caused usually by a decrease in the velocity of the water movement.

Silurian

One of the geological periods in the *Paleozoic* era, which began about 360 million years ago and lasted for about 35 million years.

Silva

The aggregate of the forest trees in an area or country.

Silviculture

The production and care of forest trees.

Simian

Refers to monkeys and apes, particularly anthropoid apes; or used as a noun especially for the latter.

Simoon

An intensely hot, dry wind of Arabian and Saharan deserts, usually carrying much sand.

Sinkhole

A hole into which water drains and passes into an underground channel, occurring usually in limestone regions.

Sinter

Deposits, mainly siliceous and calcareous (*Travertine*), formed in lakes or springs by evaporation, e.g., terraces of siliceous sinter around hot springs in Yellowstone National Park.

Sippe
A plant in the abstract as compared to the concrete plant.

Sirocco
A hot, south wind, occasionally dust laden, blowing from the Sahara in the Mediterranean region.

Site
An area delimited by fairly uniform climatic and soil conditions, essentially equivalent to *Habitat* (q. v.).

Site Index
A numerical evaluation of the quality or productivity of land, especially used in forest land where it is determined by the rate of growth in height of one or more of the tree species.

Site Quality
The capacity of a *Site* to produce vegetation, particularly timber or forage.

Skiophyte
See *Heliophobous*.

Slack
A damp hollow, or low area, among sand dunes.

Slash
The branches, cull logs, uprooted trees, and other waste material left on the ground after an area has been logged off.

Slashing
An area of forest that has been logged off and where the *Slash* remains.

Slate
A fine-grained, dense rock produced from clay or shale by compression, splitting readily into thin plates.

Sleet
(1) In the United States, frozen or partly frozen raindrops in the form of particles of clear ice. (2) In British use, snow and rain falling together.

Slick (Slick Spot)
A small area of *Alkali* or *Solonetz* (q. v.) soil.

Slime Mold
An organism in the phylum Myxomycophyta, characterized in part by a naked, fluid mass of protoplasm that can move with a flowing motion. cf. *Wasting disease.*

Slip
The downhill movement for a short distance of a mass of wet or saturated soil.

Slope
The inclination of the surface of the land from the horizontal. Level 0-3.0° (0-5%), gentle 3.0-8.5° (5-15%), moderate 8.5-16.5° (15-30%), steep 16.5-26.5° (30-50%), very steep 26.5-45° (50-100%), precipitous above 45° (over 100%).

Slough
A wet depression with deep mud. cf. *Swamp, Marsh, Bog.*

Sludge
Muddy, ooze-like sediments in a river bed, tidal flat, or similar location.

Smog
A polluted atmosphere in which products of combustion such as hydrocarbons, soot, sulfur compounds, etc., occur in detrimental concentrations for human beings and other organisms, especially during foggy weather.

Snow Density
The water content of snow expressed as a percentage by volume. In snow surveys, the ratio of the scale reading (inches of water) to the length of the core of snow in inches.

Snowfence
A fence of slats and wire or other material used to intercept drifting snow.

Snowfield
An area or mass of snow that remains throughout the summer.

Snowflush
A deposit of soil material accumulated in a mass of snow following melting of the snow.

Snow Line
A line marking the lower limit of perpetual snow.

Snow-Patch
An area in which snow melts late in the year and where *Snowflush* forms and vegetation is characteristic of such a site or is lacking.

Snow Sample
A core taken in an accumulation of snow from which the depth and density of the snow may be determined.

Snow Survey
A series of measurements of the depth and density of the accumulation of snow, usually for the purpose of determining the amount of water that is stored in the form of snow on a drainage basin, as a means of forecasting the later run-off.

Sociability
The distribution of organisms in relation to one another as individuals or as groups within a community. J. Braun-Blanquet recognizes five classes of sociability, ranging from isolated individuals to dense masses.

Social Behaviour
The activity of an animal caused by another animal or

that influences another animal; the reciprocal interactions of two or more animals.

Social Dominance

The behaviour pattern in which one or more animals dominate other individuals in the group. cf. *Peck-dominance.*

Social Facilitation.

See *Facilitation, social.*

Social Hierarchy

See *Hierarchy.*

Sociation

A vegetation type characterized by *Dominant* species in the various strata; a subdivision of the *Association* (4) (q. v.) in the Scandinavian School of Phytosociology.

Socies

According to F. E. Clements a group of one or more kinds of subdominant plants in a stage of *Succession* preceding the *Climax.* cf. *Associes, Society.*

Society

(1) A social group of individuals of one species which cooperate in their activities. (2) According to F. E. Clements a group of subdominant plants in *Climax* vegetation, cf. *Association, Socies.*

Sociology

The study of the development, composition, characteristics, and interactions of groups of organisms or communities. cf. *Plant sociology, Ecology.*

Sod

A surface layer of soil matted or bound together by roots and rhizomes of grasses and other herbs, especially by *Sod grasses.*

Sod Grass

A grass that forms a *Sod,* e.g., Kentucky bluegrass.

Softwood

The wood of a coniferous tree, e.g., pine, in contrast to *Hardwood* (q. v.).

Soil

The aggregate of weathered minerals and decaying organic material that covers the earth in a thin layer in which plants grow.

Soil Creep

The very slow movement of surface soil down a slope.

Soil Erosion

The loosening and movement of particles of soil from the surface of the land by wind or flowing water, including *Accelerated erosion* and *Normal erosion.* cf. *Gully erosion, Rill erosion, Sheet erosion, Splash erosion.*

Soil Horizon

A layer of soil with characteristics resulting from soil-building processes. See *A, B, C horizons, Podzolization.*

Soiling

The feeding of livestock with mowed, fresh forage such as bromegrass or legumes, in contrast to their grazing on a pasture.

Soil Productivity

The capacity of a soil to produce plant growth because of its chemical, physical, and biological properties.

Soil Profile

A vertical section of the soil from the surface through all its horizons into the parent material. cf. *Soil horizon.*

Soil Reaction

The acidity or alkalinity of the soil usually expressed as *pH* (q. v.).

Soil Structure

The arrangement of particles in the soil, e.g., single grain, granular, columnar.

Soil Texture

The relative proportions of the various sizes of mineral particles (gravel, sand, silt, clay) in the soil. cf. *Silt.*

Soil Type

An area of soil which is relatively uniform in profile characteristics and in texture of the surface soil, a subdivision of a soil series, e.g., Cecil sandy loam and Cecil clay loam are soil types in the Cecil series.

Solar Constant

The energy received from the sun above the upper limit of the atmosphere, equal to 1.94 gram-calories per minute per square centimeter.

Solarization

The inhibiting effect of extremely high light intensities on *Photosynthesis.*

Solifluction

The flow of saturated soil upon an impermeable layer or on frozen ground, especially under conditions of alternate freezing and thawing.

Solonchalk

A type of soil that has a high concentration of soluble salts in relation to other soils, usually light-colored, "white alkali." cf. *Saline soil.*

Solonetz

A type of soil in which the surface horizons of varying friability are underlain by dark-colored, hard soil which is usually highly alkaline and columnar in structure, "black alkali." cf. *Alkali soil, Solonchalk.*

Solonization

The process of soil formation in semiarid and arid climates where *Saline soil* (q. v.) or *Solonetz* is formed. cf. *Podzolization, Calcification.*

Solstice

The time of the year when the sun is above the point which is farthest north or south of the equator, in the northern hemisphere the summer solstice is about June 21, the winter solstice about December 22.

Solum

The upper part of the *Soil profile* (q. v.) above the parent material, usually the *A* and *B horizons;* often considered the true soil because of its development by soil-building forces. cf. *Soil.*

Soma

The cells of an organism exclusive of those concerned with *Sexual reproduction.*

Somatic

Refers to the *Soma,* or the non-reproductive parts of an organism.

Sonoran Life Zone

The part of the *Austral life zone* (q. v.) lying west of the 100th meridian, divided into *Transition, Upper Sonoran,* and *Lower Sonoran* zones.

Spat

A juvenile form of a bivalve mollusk such as the oyster.

Spawn

(1) The eggs of frogs, fishes, oysters and other aquatic animals. (2) The *Mycelium* (q. v.) of certain fungi especially of the mushroom in which it is used for propagation.

Spay

To remove the ovaries from a female animal.

Specialized

Refers to an organism, or part thereof, that is adapted to a particular kind of life or to a certain combination of environmental conditions; more limited than an unspecialized organism.

Speciation

The processes in evolution by which new species are formed. cf. *Mutation, Natural selection, Subspeciation.*

Species

A unit of classification of plants and animals, consisting of the largest and most inclusive array of sexually reproducing and cross-fertilizing individuals which share a common gene pool; the most inclusive *Mendelian population* (q. v.), e.g., the white pine *(Pinus strobus)* and ponderosa pine *(Pinus ponderosa)* are two species in the genus *Pinus.* cf. *Jordanon, Ecospecies, Coenospecies, Taxon, Syngameon, Superspecies.*

Species-area Curve

A graph showing the number of species on the vertical axis and the area of the sampling-unit or *Quadrat* on the horizontal axis; used to determine the most suitable area of quadrat to use in sampling vegetation.

Specific Gravity (Soils)

The ratio of the weight of a given volume of soil, pore space excluded, to the weight of an equal volume of water; the average specific gravity of tilled surface soil is about 2.65.

Specificity

The limitation of an organism to restricted, definite set of environmental conditions, a single kind of food plant or animal host, or other set of circumstances.

Spectrum, Biological

See *Biological spectrum.*

Speleology
The study of the conditions and the life in caves.

Sperm
The male sex cell or *Gamete.*

Spermatogenesis
The formation of sperms in an organism.

Spermatophytes
The seed-bearing plants, Spermatophyta, a section of the subphylum Pteropsida, phylum Tracheophyta. It includes the *Gymnosperms* and the *Angiosperms* (q. v.).

Spermatozoon
A highly motile *Sperm* occurring in animals.

Sphagniherbosa
Plant communities with abundance of *Sphagnum* and with peat in the substratum. cf. *Bog.*

Sphagnum
A genus similar to the true mosses, in the subclass Sphagnobrya, class Musci, phylum Bryophyta; usually occurring in bogs.

Sphagnum Bog
A kind of community characterized by the presence, and often the abundance, of *Sphagnum,* acid substrata, and the accumulation of peat. cf. *Bog.*

Spikelet
One of the main parts of the inflorescence of a grass or a sedge, containing one or more flowers (*Florets*) and associated bracts or scales.

Spillway
A passageway for the escape of excess water around a dam.

Spinney (Spinny)
A copse or small grove.

Spiracle
One of the external openings of the *Trachea* (q. v.) of most terrestrial *Arthropods.*

Spirochete (Spirochaete)
A microorganism which moves by undulating its body (not by cilia), parasitic or free-living, classified usually with the bacteria, e.g., the organism causing syphilis.

Spit
A long, narrow strip of land extending into the sea, attached to the mainland at one end.

Splash erosion
The direct effect of the impact of rain drops on the ground surface or on a thin film of water causing detachment of soil particles which are then readily available for washing away. cf. *Erosion, Sheet erosion, gully erosion.*

Sponge
An animal in the phylum Porifera.

Spontaneous Generation
The belief that organisms, even complicated ones, originated directly from non-living substances. cf. *Biogenesis.*

Sporangium
A case-like structure in plants in which spores are produced.

Spore
An asexual, *Haploid* (q. v.), one- or few-celled, reproductive body produced by organisms. cf. *Spore-mother cell.*

Spore-mother Cell
A *Diploid* (q. v.) cell in plants that gives rise to four *Haploid* spores.

Sporophyll
A leaf or leaf-like structure or scale that produces one or more *Sporangia,* e.g., leaves of many ferns, a *Stamen.*

Sporophyte

The part, or asexual generation, of the life cycle of plants in which the cells contain the *Diploid* (or *Polyploid*) number of chromosomes, begins with *Fertilization*, produces *Spores*, e.g., a flowering plant, a fern.

Sport

A vegetative or *Somatic Mutation* (q. v.) in an organism, e.g., a shoot differing from other shoots arising from a bud on a plant.

Sporulation

The rapid formation of *Spores* by fission as in many bacteria, molds, algae, and protozoons.

Spread

The combined results of dispersal and of the establishment of the individual and then the species in a new place. cf. *Dispersal, Establishment.*

Sprigging

The planting of a part of the stem and root system of a grass.

Spring Overturn

The mixing of water in lakes after the ice melts, resulting in a uniform temperature from the surface to the bottom. Another mixing occurs in the autumn, the *Fall overturn.*

Springtail

See *Collemboion.*

Spring Wood

The portion of the annual woody growth of a tree or a shrub that is formed in the early part of the growing season; it is more porous than the *Summer wood* (q. v.).

Sprout

The first growth or shoot from a seed, root, or other plant part; or a tree that has grown from a stump or root.

Square-foot Method

A method used to determine the species composition and the cover of range vegetation by means of systematically located sample areas one square foot in area.

Stabilization

(1) The state in the interrelationships of organisms in which integration and adjustment between the organisms and between them and the prevailing environment is being attained, maximum stabilization occurs in climax communities usually. (2) In oceanography the condition in a mass of water in which a density gradient has become established such as when a *Thermocline* occurs.

Stamen

The part of a flower that produces pollen, consisting of an *Anther* (contains the pollen) and a filament (the stalk).

Staminate

Refers to a flower that bears stamens.

Stand

A general term for an aggregation of plants with more or less uniformity in *Physiognomy*, composition, and habitat conditions; a local example of a *Community-type* or *Association* (q. v.).

Standing Crop

The total amount of the *Biomass* (q. v.) of organisms of one or more species within an area. cf. *Productivity, Yield.*

Standort

The influence in the aggregate of all factors (climatic, edaphic, biotic, orographic) upon a geographically delimited locality. cf. *Habitat.*

Stand Table

A listing of species that occur in a stand, including data

on characteristics such as *Cover, Vitality,* and *Frequency.* cf. *Association table.*

Station

A particular location comprising a stand, part of a stand, or a locality. cf. *Habitat, Standort.*

Steady State

See *Homeostasis.*

Stele

The central part of the stem or root of plants, includes the pericycle, *Phloem, Xylem,* and pith when present.

Steno-

A prefix denoting a narrow range of *Ecological amplitude* (q. v.) of an organism, e.g., stenothermal refers to temperature, stenophagous to variety in the diet, stenoky to number of factors, stenohaline to salinity, stenohydric to water. cf. *Euroky.*

Steppe

An extensive area of natural, dry grassland; usually used in reference to grasslands in southwestern Asia and southeastern Europe. cf. *Prairie, Pampas.*

Stereotaxis

See *Thigmotaxis.*

Stereotropism

The growth of a plant organ in response to contact with an object, e.g., tendrils of vines coiling around a stem.

Sterility

The lack of ability of an organism to carry on *Sexual reproduction.*

Stigma

The upper part of the pistil (carpel) of a flower, receives the pollen and aids in its germination. cf. *Pollination.*

Stilling Basin

An excavation or structure below a waterfall or rapids that reduces the velocity and turbulence of the current.

Stimulus

An influence that causes a response in an organism or in a part of it.

Stock

(1) The parts of a plant, usually a portion of the stem and the root system, to which a *Scion* is grafted. (2) Livestock. (3) See *Standing crop*.

Stocking

Placing animals such as deer or domestic livestock on an area of vegetation.

Stolon

A horizontal stem on the surface of the ground where it propagates vegetatively by forming new *Shoots* and roots at the nodes, e.g., Bermuda grass. cf. *Runner, Rhizome*.

Stoma (Stomate)

A minute pore and two surrounding guard cells occurring in the epidermis of leaves, young stems and fruits, and other organs, through which diffusion of gases occurs.

Stool

The base of a plant from which shoots arise, or the base including the shoots. cf. *Tiller*.

Stover

The dry, cured stems and leaves of grain crops such as corn and sorghum after the removal of the grain. cf. *Fodder, Forage, Feed*.

Strain

A group of organisms having distinctive attributes and a common lineage which differs from other groups, but which is not sufficiently distinct to form a breed or *Variety*.

Strand

(1) The area of bare beach above the level of high water, which is subject to the action of wind. (2) The intertidal portion of a beach.

Stratification

See *Layering, Thermal stratification.*

Stratified Sampling

In sampling vegetation or a geographic complex the separation into types or blocks in order to secure the maximum degree of *Homogeneity* in the area to be sampled.

Stratosphere

The upper region of the atmosphere beginning about six miles above the surface of the earth, in which water-vapor clouds do not form and where no marked changes in temperature take place as the altitude increases.

Stratum

See *Layer.*

Stream

A general term for water flowing in one direction such as a rill, rivulet, brook, creek, and river.

Streptomycin

An *Antibiotic* (q. v.) produced by the mold, *Streptomyces griseus.*

Stress

(1) Systemic. According to Selge the condition of an animal in which large parts of the body deviate from their normal resting state, either because of their activity or because of an injury. (2) The total energy with which water is held in the soil.

Stressor

A stimulus causing systemic stress.

Stridulation

The making of shrill sounds by certain insects such as crickets by rubbing one organ against another.

Strip Cropping

The growing of crops in narrow fields or strips so that wind and water erosion is reduced or prevented. cf. *Buffer strips.*

Strip Survey

The use of continuous narrow strips as sampling units, especially in forestry.

Strobilus (Strobile)

A cluster of *Sporophylls;* the cone of conifers in which seeds or pollen grains are produced.

Structure

(1) An expression of the composition, abundance, spacing, and other attributes of plants in a community. cf. *Layering, Life-form.* (2) The composition of a population with reference to age-classes, or to some other criterion. (3) See *Soil structure.*

Struggle for Existence

Refers to the processes used by an organism to maintain life and to reproduce, especially in an unfavorable environment or where *Competition* (q. v.) is severe. cf. *Natural selection.*

Stubble

The lower parts of plants that remain after the tops have been removed in harvesting operations; may also be applied to the parts left ungrazed on range or pasture.

Stubble Crop

(1) A crop that is produced from *Stubble* of the previous season. (2) A crop sowed in the grain *Stubble* after the grain

crop such as wheat has been harvested; for the purpose of plowing it under the following spring to increase organic matter in the soil.

Stubble Mulch

The residues of a crop left on the soil surface as a mulch to prevent or reduce erosion when preparing the land for planting another crop.

Stumpage

The value of timber as it stands in a forest; the uncut timber.

Style

The portion of the pistil between the stigma and the ovary in a flower.

Subalpine

Refers to the region or zone in mountains below the treeless *Alpine* (q. v.) region, characterized in North America by coniferous forests, especially spruce and fir.

Subarctic

(1) Refers to the region south of the *Arctic* (q. v.) region and includes the northern part of the region south of the geographical timber line. (2) Boreal.

Sub-boreal Period

The climatic period from about 2500 B. C. to about 700 B. C. according to Blytt and Sernander, a period drier than the preceding *Atlantic* (5500-2500 B. C.) and the following Subatlantic (700 B.C. to the present) periods. cf. *Boreal period.*

Subclimax

A subfinal stage in *Succession* in which further development is inhibited because of the influence of some factor other than the climatic factors. cf. *Proclimax, Serclimax.*

Subdominant

A species in a community that exerts much less *Dominance* (q. v.) than the *Dominant* species.

Suberin

The waxy material found in walls of chiefly cork cells in plants.

Suberization

The process of *Suberin* formation in plants.

Subhumid

Refers to climatic regions where the moisture conditions range from 20 inches in the cool parts to 60 inches in the hot parts; and where the natural vegetation consists chiefly of tall grasses, and where many kinds of crops can be grown without irrigation, or dry farming procedures.

Subinfluent

An organism that has less effect than an *Influent* in a community and is present usually for only a part of the year.

Subirrigation

The control of the water table so as to raise it near or into the root zone.

Sublittoral

The lower division, at a depth from about 40 or 60 meters to about 200 meters in the sea, of the *Neritic* or *Benthic zone,* below the *Littoral* division. These terms apply also in a general way to lakes.

Subpolar Region

Approximately the region south of the *Tundra,* occupied by *Boreal* forest.

Subsequent Reproduction

Trees which have grown up in openings in the forest

or under the canopy following cutting or after regeneration operations have been started.

Subsere

See *Secondary succession.*

Subsoil

Approximately the *B horizon* in soils that have distinct profiles; where the profile development is weak the subsoil is below the plowed soil, or its equivalent, in which roots normally grow, a vague term.

Subsoiling

Tillage of the *Subsoil* (q. v.) or the soil below the normal depth of plowing. cf. *Chiseling.*

Subspeciation

The formation within a species of populations that differ consistently one from another in *Genotypic* constitution and in the resulting *Phenotypes.* Isolation of such *Subspecies* may in time give rise to new *Species.* cf. *Speciation.*

Subspecies

A *Taxon* of distinct, geographically separated complexes of genes, immediately below *Species* and above *Variety* (if varieties are recognized in a species), sometimes considered as synonymous with variety, or as an incipient species.

Substitute Species

See *Vicariation.*

Substratum

(1) The base or substance upon which an organism is growing. (2) A vague term for the *C horizon* (q. v.).

Subtropical

Refers to the region between the *Tropics* and the *Temperate zone,* with distinct summer and winter seasons and with greater heat than the Temperate zone.

Succession (Ecological)

The replacement of one kind of *Community* by another kind; the progressive changes in vegetation and in animal life which may culminate in the *Climax* (q. v.). cf. *Allogenic, Autogenic, Primary succession, Secondary succession, Sere.*

Succulence

The condition of a plant that contains much tissue rich in cell sap and is therefore fleshy or juicy, e. g., cactus.

Sucker

(1) In some animals an organ of attachment and also often used for the absorption of food. (2) See *Haustorium.* (3) In many plants a shoot arising from the lower parts of the stem or from the root. cf. *Tiller, Sprout.*

Sudd

An extensive *Marsh* type of vegetation characterized by the *Dominance* of papyrus (*Cyperus papyrus)* along the upper White Nile River, large masses of which may break loose and float down the river.

Suffrutescent

Refers to *Perennial* plants that normally are somewhat woody at the base so they do not die down to the ground each year.

Suffruticose

Refers to *Perennial* plants that are distinctly woody at the base, herbaceous above (*Undershrubs*), intermediate to *Suffrutescent and Fruticose* (q. v.).

Sulfofication

The *Mineralization* (q. v.) of organic compounds in dead remains of plants and animals to inorganic compounds containing sulfur such as calcium sulfate which can again be absorbed by plants.

Summation Temperature

See *Temperature summation.*

Summer Fallow

The cultivation of a field in which crops have not been planted in order to control weeds and to accumulate soil moisture for the growth of a crop subsequently. cf. *Fallow.*

Summer Wood

The less porous and harder portion of the *Xylem* (q. v.) of a growth layer in woody plants produced in the latter part of the growing season. cf. *Spring wood.*

Sun Plant

A plant that grows well in full sunlight. cf. *Heliophyllous.*

Sunscald

Death of tissues of a plant caused by high temperature and loss of water in organs exposed to bright sunshine.

Sunspot Cycle

The alternation in occurrence of a period of numerous spots on the surface of the sun and a period with fewer spots; one cycle averages about 11 years.

Superorganism

See *Epiorganism.*

Superparasite

A secondary *Parasite,* i. e., a parasite using another as its *Host.*

Supersonic

Refers to vibrations exceeding 20,000 per second, not audible to the human ear.

Superspecies

A group of related Species that are geographically isolated; without any implication of natural *Hybridization* among them. cf. *Syngameon.*

Supralittoral (Zone)
The portion of the shore immediately adjacent to the tidal zone.

Supraneuston
Collectively, the minute organisms associated with the upper surface of the film of water in lakes, streams, etc. e. g., the water-strider.

Supraorganism
See *Epiorganism.*

Surface Run-off
See *Run-off.*

Surface Soil
The upper part of cultivated soil, usually stirred during tillage operations, or the equivalent depth of 5 to 8 inches in non-cultivated soils.

Surplus Stock
The portion of the population of game animals or fish at the time of harvest that are in excess of the number needed to maintain an adequate breeding stock.

Survival Potential
The capacity of an organism to survive in a given environment.

Swale
An area of low, wet land; a low meadow.

Swamp
A land area containing excessive water much of the year and covered with dense, native vegetation that includes trees; but the term is used with various meanings. cf. *Marsh, Bog.*

Sward
An area of grassland, especially one composed of sod-grasses. cf. *Turf.*

Swarm

A dense aggregation of minute aquatic organisms, or of certain insects such as bees and midges.

Sweep-net Method

A technique for determining an evaluation of the density of insects and other invertebrates in an area by making a certain number of swings of a standard entomological sweep net.

Sweepstakes Bridge

The accidental transportation of organisms across a barrier from one area to another, usually where no land connection occurs, e. g., the migration of a few kinds of animals from Africa to Madagascar. cf. *Filter bridge, Corridor.*

Symbiont

In a broad sense an organism that lives in close association with another. cf. *Symbiosis.*

Symbiosis

In a broad sense the living together of two or more organisms of different species; including *Parasitism, Mutualism,* and *Commensalism* (q. v.). cf. *Coaction.* In a narrow sense synonymous with mutualism.

Symmetry

The condition of similarity in form or structure in the parts of an organism on each side of an axis dividing it. cf. *Bilateral symmetry, Zygomorphy.*

Sympatric

Refers to the origin or area of occupation of two or more closely related species in the same geographical area. cf. *Allopatric.*

Synapse

One of the places in the nervous system of animals where

nerves touch one another and where stimuli are transmitted from one nerve cell to another.

Synapsis

The pairing of *Homologous chromosomes* (q. v.) in early stages of *Meiosis* (q. v.).

Synchorology

The branch of *Plant sociology* dealing with the occurrence and distribution of communities. cf. *Plant geography.*

Syncline

A geological structure or fold formed by strata from opposite sides dipping downward toward a common line. cf. *Anticline.*

Synconium

A kind of fleshy fruit in which the seeds are produced on the inner surface of the concave or hollow receptacle, e. g., fig.

Syncytium

In certain animals a mass of cytoplasm containing many nuclei within a single plasma membrane. cf. *Coenocyte, Plasmodium.*

Syndactylism

The condition in which two or more digits are at least partly joined.

Synecology

The study of the environmental relations of communities, a branch of *Plant sociology.*

Synergism

The total activity of separate agents such as various drugs producing an effect which may be greater than the sum of the effects of the individual agents.

Syngameon

The sum total of species linked by frequent or occasional *Hybridization* in nature; a hybridizing group of species; the most inclusive interbreeding population.

Syngamy

The fusion of *Gametes;* the *Fertilization* of an egg by a sperm.

Syngenetics

The branch of *Plant sociology* dealing with the origin and development of communities. cf. *Succession, Community dynamics.*

Synthesis Table

See *Association table.*

Synusia

An aggregation of plants belonging to the same *Life-form* having similar environmental requirements and occurring in a similar *Habitat,* e. g., a layer of moss plants, a group of floating herbs such as water lilies.

Systematic Plant Sociology

The branch of *Plant sociology* that deals with the delimitation and description of communities, followed by grouping them into categories such as *Sociation, Association, Alliance, Order,* and *Class.*

Systematic Sampling

A method of sampling in which the samples are distributed in a regular manner so that the sampling units will be located as uniformly as possible over the area under study.

Systematics

The science of classification; including the description, naming, and grouping of organisms in categories such as species, genus, family, order, and class; with especial consideration of evolutionary relationships.

Systemic

Refers to the entire body of an organism, e. g., the whole body of an organism being affected by a disease.

T

Tableland

A broad, elevated area of land bounded by steep slopes or cliffs. cf. *Mesa.*

Taiga

The open forest, usually coniferous, adjacent to the arctic *Tundra* (q. v.) cf. *Boreal Forest.*

Tailings

Accumulations of coarse rock debris from which the finer materials have been removed during mining operations.

Tallgrass Prairie

See *True prairie.*

Talus

Accumulations of rock fragments below steep slopes or cliffs, caused by the effect of gravity.

Tame Pasture

An area of land once cultivated and seeded to cultivated plants, used for grazing. cf. *Range, Ley.*

Tank, Earth

A structure made by an excavation and an earthen dam across a drainage course for the purpose of impounding drinking water for livestock.

Tapeworm

A flatworm in class Cestoda, phylum Platyhelminthes, parasitic in the adult stage in the intestines of *Vertebrates.*

Taproot System

A root system in plants characterized by a large primary root (the taproot) that extends deep into the soil and has many smaller branches, e. g. alfalfa. cf. *Fibrous root system.*

Tarn

A small lake or pool in the mountains.

Taungya

An area cleared of vegetation and undergoing *Secondary Succession* in Burma.

Taxis

Movement of an organism directly towards or away from a stimulus, e. g. *Phototaxis,* (q. v.)

Taxocline

A series of gradations in taxonomic status of organisms in which hybridization is involved.

Taxon

Any taxonomic category, e. g., species, genus, variety.

Taxonomy

The science of classification of organisms; the arrangement of organisms into systematic groupings such as *Species, Genus, Family,* and *Order.* cf. *Systematics.*

Tectonic

Refers to processes that cause the formation of features of the earth's crust, e. g., upwarping. cf. *Isostasy.*

Teleology

The belief that the processes of nature are directed towards some end or goal such as plants store starch for the purpose of surviving.

Telotaxis

The direct orientation of an organism to the gradient of a stimulus, known only in response to light. cf. *Taxis, Tropotaxis.*

Temperate Zone

The portions of the earth in the northern and southern hemispheres between the *Tropics* (q. v.) and the polar circles 23°27′ from the poles. cf. *Frigid zone.*

Temperature Coefficient

See Q_{10}.

Temperature, Effective

The temperature above a certain minimum, at which physiological processes such as growth of an organism are active, considered 5°C (41°F.) for many plants.

Temperature Inversion

See *Inversion, temperature.*

Temperatures, Cardinal

The minimum, optimum, and maximum temperatures for the growth of an organism or organ, or for a process or activity.

Temperature Summation

The summing of effective temperatures or *Day-degrees* (q. v.) for a period of time or for the length of time required for the development of an organism or organ. See *Aliquote.*

Temperature Zero

The temperature below which a number of physiological processes of an organism cease or are carried on at a very slow rate. cf. *Temperature, effective.*

Temporary Pasture

A pasture used for grazing for only a short period, usually composed of annual plants. cf. *Tame pasture.*

Tendon

A band of dense, fibrous tissue connecting a muscle to some other part, usually a bone, in an animal.

Tendril

A stem, leaf, leaflet, or stipule of a plant, modified into a slender structure that coils around an object thus giving support to the plant bearing the tendril, e. g., pea vines, clematis.

Tensiometer

An instrument for measuring the tension with which water is held in the soil. cf. *Stress.*

Tentacle

A slender, flexible organ, usually tactile, attached to the head of many kinds of animals such as insects, jellyfish, and snails; also the hair-like structures on the insectivorous sundew plant which traps insects.

Teratology

The study that deals with monstrosities and malformations in organisms, especially in man.

Termitarium

A mound constructed and inhabited by termites.

Termites

Animals in the order Isoptera (white ants), resembling true ants, forming large, complex colonies with a highly developed social system, occurring especially in the *Tropics.*

Termiticole

An organism inhabiting a termite nest.

Termitiphile

An organism living with termites in their galleries.

Terrace

(1) Flat or undulating land usually with a steep face bordering a stream, lake, or sea cf. *Floodplain*. (2) An embankment of earth built across a slope to control *Run-off* and reduce erosion.

Terrestrial

Refers to the land.

Terrigenous

Refers to deposits derived from the land. cf. *Allochthonous, Autochthonous.*

Terriherbosa

Herbaceous types of vegetation on dry land, e. g., *Steppe, Prairie.*

Territoriality

The behaviour of an animal when it defends an area from intruders, e. g., various birds and fishes.

Territory

(1) The area occupied by an individual or group of organisms. (2) The area which an animal defends against intruders. cf. *Home range.*

Terron

An earthen construction made of bricks that have been cut directly from the natural sod of sedge meadows and dried in the sun. cf. *Adobe.*

Tertiary

The first of two geological periods in the Cenozoic era comprising the *Paleocene, Eocene, Oligocene, Miocene,* and *Pliocene* epochs; in order from the oldest to the most recent.

Testa

The outer covering or coat of seeds.

Tetraploid

An organism or part of one having four sets of *chromosomes* in its nuclei. cf. *Haploid, Diploid, Polyploid.*

Texture (Soil)

The property of the composition of soil that deals with the relative proportions of various sizes of separates or mineral particles including clay, silt, sand, and gravel. cf. *Structure.*

Thallophyte

A plant in any one of the phyla of algae and fungi, formerly classified in the division Thallophyta.

Thallus

A plant body that is not differentiated into leaves, stems, and roots; one- to many-celled, e. g., *Thallophytes.*

Thermal Constants

The sum of *Day-degrees* (q. v.) of temperature that is required for a plant to mature after planting. cf. *Temperature summation.*

Thermal Stratification

The condition of a body of water in which the successive horizontal layers have different temperatures, each layer more or less sharply differentiated from the adjacent ones, the warmest at the top. cf. *Epilimnion, Hypolimnion, Inverse stratification, Thermocline.*

Thermal Zone (Belt)

A well defined area or zone, occurring on some mountainsides, in which the vegetation is exceptionally free from frost in the spring and fall.

Thermocline

The layer in a thermally stratified body of water within which the temperature decreases rapidly with increasing depth usually at a rate greater than 1°C. per meter of depth. cf. *Thermal stratification.*

Thermodynamics, Laws of

(1) Energy and work are transformable from one to another kind, e. g., sunlight to chemical energy. (2) Spontaneous transformation of energy is accompanied by dispersal of a part into non-available heat such as in respiration. (3) The absolute zero temperature is not attainable.

Thermogenic

The production of heat as occurs in an organism during respiration.

Thermogram

The continuous record of temperature made by a *Thermograph.*

Thermograph

A self-recording thermometer.

Thermonasty

The response of an organism to a general diffuse change in temperature, e.g., the opening of flowers as the temperature rises.

Thermoperiodism

The effects of the alternation of temperature such as occurs during day and night alternations upon organisms.

Thermophilous

Refers to organism that grows well in high temperatures, e.g., bacteria in hot springs.

Thermotaxis

The movement of an organism toward heat or cold as a stimulus.

Therophyte

One of the classes of life-forms of *Raunkiaer* that includes the annual plants.

Thiamine

A vitamin (B_1) required by numerous organisms, but formed only in green plants and in some microorganisms.

Thicket

Vegetation that is dominated by a dense growth of small trees and shrubs.

Thigmotaxis

The movement of an organism to secure close contact with an object. syn. *Stereotaxis.*

Thigmotropism

The response of a plant or a portion of it to a contact stimulus, e.g., a tendril growing around a stem. syn. *Stereotropism.*

Thorax

(1) The part of the body in higher *Vertebrates,* containing the heart and lungs. (2) The middle portion of the body of insects, bearing the legs and wings (when present).

Thorn

A stiff, pointed, modified branch in plants such as the hawthorn.

Thorn Forest

A vegetation type in the *Tropics* or *Subtropics* consisting mostly of thorny trees, shrubs, and vines; *Xerophytic* in aspect, and subject to long droughts.

Threadworm

See *Nematode.*

Threshold

The duration or intensity of a stimulus that is required to produce response in an organism.

Thrombosis

The coagulation or clotting, as of blood, in the vascular and lymphatic systems of living animals.

Thyroid

One of the *Endocrine glands* (q. v.) found in *Vertebrates,* secretes a *Hormone* containing iodine.

Thyroxine

The substance containing iodine produced by the *Thyroid gland.*

Tick

An animal in the order Acarina, class Arachnida, which sucks blood, e.g., the fever tick on cattle. Also used for some parasitic dipterous insects such as the sheep tick.

Tidal Flat

An essentially barren, nearly flat, muddy area, periodically covered by tides, the lower parts daily, the higher parts only during exceptionally high tides.

Tidal Marsh

A low flat marshland that is intersected by channels and tidal sloughs, usually covered by high tides, with vegetation consisting of rushes, grasses, and other low, salt-tolerant plants.

Tidal Zone

The area of a shore between the levels of high and low tides. syn. *Intertidal zone.*

Till

An unstratified deposit of gravel, boulders, sand, and finer materials which has been transported by a glacier. cf. *Drift, Colluvium.*

Tillage

The operations such as plowing, harrowing, and disking that are used in cultivating the soil in order to make conditions more suitable for the growth of crop plants.

Tiller

A *Shoot* arising from the base of a plant as in wheat and other grasses.

Till Plain

A more or less level land area covered with glacial *Till.*

Tilth, Soil

The physical condition, particularly *Structure,* of soil with reference to favorableness for growth of crops; characterized by friability, high degree of non-capillary porosity, and stable granular structure. Soil in poor tilth is non-friable, hard, non-aggregated, and difficult to cultivate properly.

Timberline

This term usually denotes the upper limit of tree growth in mountains or poleward in latitude. cf. *Tree limit.*

Tissue

A organized, usually compact group of cells that have similar structure and function, e.g., cork tissue in plants, bone tissue in animals.

Tissue Culture

The growth in a suitable medium of a portion of tissue separated from a plant or animal body.

Tolerance

The capacity of an organism to live under a given set of conditions within its range of *Ecological amplitude* (q. v.), between the *Maximum* and the *Minimum* (q. v.) (the limits of tolerance). cf. *Preferendum.*

Topography

A general term to include characteristics of the ground

surface such as plains, hills, and mountains; degree of *Relief,* steepness of slopes; and other physiographic features.

Topsoil

The uppermost portion of the soil, often considered the layer six or seven inches in thickness, which is richer in organic material and lighter in texture than the material below. In uniform material the topsoil includes the layer that is usually plowed up. cf. *Subsoil.*

Tornado

(1) A violent vortex, with a diameter usually of about 0.25 mile, in the atmosphere, accompanied by a pendulous, more or less funnel-shaped cloud. (2) In West Africa a violent thundersquall.

Torrential

Refers to the rapid and violent movement of materials such as water in a stream, heavy rainfall, sliding of gravel, etc.; or to organisms living in swift streams.

Torrid Zone

See Tropics.

Total Digestible Nutrients (T. D. N.)

The standard evaluation of the digestibility of materials in the feed of livestock, including proteins, fats, nitrogen-free extract, and fiber content.

Total Estimate

The combined estimates of abundance and cover characteristics of vegetation; used commonly in England.

Toxin

A poison produced by an organism.

Trace Element

See *Essential element.*

Tracer

A *Radioisotope isotope* (q. v.) used to follow the course of or to determine the location of a normal element in an organism.

Trachea

(1) A *Vessel* in the *Xylem* (q. v.) in plants. (2) The tube from the throat to the lungs in *Vertebrates.* (3) One of the small tubes that conduct air in the bodies of most *Arthropods,* especially in insects, opening to the exterior through *Spiracles* (q. v.).

Tracheid

A long, thick-walled cell, without perforations, with tapering ends, that conducts water and gives support, located in the *Xylem* of plants, especially *Conifers.*

Tracheophyte

A member of the Tracheophyta (vascular plants) comprising four subphyla; Psilopsida (the most primitive), Lycopsida (the clubmosses), Sphenopsida (the horsetails), and the Pteropsida (the ferns, conifers and their allies, and the flowering plants).

Trade Winds

Winds that blow regularly from subtropical areas of high pressure towards areas of low pressure along the equator, in the northern hemisphere from the northeast, in the southern hemisphere from the southeast; important in producing ocean currents.

Transad

A species, or closely related species, that exist on both sides of a barrier and consequently must have extended across it at one time.

Transect

A long, narrow sample area, or a line, used for analyzing

vegetation; essentially a cross section of the vegetation. cf. *Line-intercept method, Quadrat.*

Transformation

The change of one type of bacterium to another type, as when *DNA* from Type I of *Pneumococcus* is transferred to Type II it replaces some of the *Chromatin* in the latter whose characteristics such as resistance to *Penicillin* may thus be modified. Genetic information is thus transferred.

Transgressive

Species that regularly occur in an upper layer in a community found also in a lower stratum.

Transhumance

The periodic and alternating movement of livestock between two regions that differ in climate.

Transient Species

A species that migrates through a locality without breeding or over-wintering.

Transition Life-Zone

The northern part of the *Austral life-zone* (q. v.).

Translocation

(1) The movement of materials in solution from one part of a plant to another part. cf. *Xylem, Phloem.* (2) The separation of a part of a *Chromosome* and its attachment to another one.

Transpiration

The loss of water in vapor form from a plant, mostly through the stomata and lenticels. cf. *Stoma, Lenticel.*

Transpiration Coefficient

The ratio of the oven-dry material produced by a plant during a more or less extended period of time (usually the entire growth period) to the total amount of water transpired during the same period.

Transpiration Efficiency
The amount in grams of dry substance produced by a plant for every kilogram of water transpired.

Transplant
A seedling, or young plant, that has been moved from one location to another; in forestry practice a seedling that has been transplanted one or more times in the nursery.

Transportation (in Soil Erosion)
The movement of detached particles or masses of soil across the land or through the air by wind, water, or gravity.

Trap Line
A series of traps arranged in a more or less linear arrangement to secure a sample of the mammals in an area, or for securing animals for their fur, or some other purpose.

Traumatic
Refers to a shock or a wound, or the resulting condition in an organism.

Travertine
A calcareous, concretionary limestone that has been formed in water. cf. *Sinter, Tufa.*

Tree
A woody plant that has a single main stem and commonly more than eight or ten feet tall. cf. *Shrub.*

Tree Limit (Line)
The altitude in mountains, or the southern or northern latitude, at which only isolated trees grow and beyond which only stunted forms, *Krummholz* (q. v.), or *Tundra* (q. v.) occur. cf. *Timberline.*

Tremotode
cf. *Fluke.*

Triassic

The oldest geologic period in the *Mesozoic* era; it began about 205 million years ago and lasted for about 40 million years.

Tribe

A group of plants of related genera, a division of the *Family*, e.g., the tribe Festuceae includes the genera *Poa, Festuca, Bromus,* and others.

Triploid

An organism or one of its parts that has three times the *Haploid* set of *Chromosomes* in the nucleus. cf. *Polyploid.*

Trisomic

Refers to an organism that has one more *Chromosome* than the *Diploid* (q. v.) number; occurs in barley, and peas.

Tristat Method

A method of photographing the same area at successive periods of time by permanently marking the spot where each leg of the tripod is set.

Trophic (-trophy)

Refers to nutrition.

Trophic Level

One of the parts in a nutritive series in an *Ecosystem* (q. v.) in which a group of organisms in a certain stage in the *Food chain* secures food in the same general manner. The first or lowest trophic level consists of *Producers* (green plants), the second level of *Herbivores,* the third level of primary *Carnivores,* the fourth level of secondary Carnivores. Bacteria and fungi are organisms in the *Decomposer* trophic level.

Trophobiosis

A type of association of species involving aphids and coccids with ants. cf. *Myrmecophilous.*

Trophollaxis
The exchange of food, or the interchange of a stimulus response concerning food, between animals, especially in the social insects.

Tropical Cyclone
See *Hurricane.*

Tropical Life Zone
The portion of Central America south of the *Austral life zone* (q. v.), bounded on the north by an accumulation of temperatures during the growing season above 43°F. of 26,000°F. cf. *Life zone.*

Tropics
The Tropic of Cancer, 23°27′ north latitude, and the Tropic of Capricorn, 23°27′ south latitude; or the region between these parallels.

Tropism
The curvature response of an organ to a stimulus, e.g., a stem growing towards a source of light and roots away from the light.

Tropoparasite
An *Obligate parasite* (q. v.) that regularly lives as a non-parasite during part of its life cycle.

Tropopause
The uppermost portion of the *Troposphere* (q. v.).

Tropophyte
A plant that can live under moist conditions part of the year and under dry conditions during another part, e.g., woody plants that lose their leaves during the dry parts of the year or during winter.

Troposphere
The part of the atmosphere extending upward about

six miles to the *Stratosphere* (q. v.), in which clouds of moisture form and the temperature decreases with increasing altitude.

Tropotaxis

The direct orientation of an organism in response to two lights that are moving toward or away from the midpoint between them. cf. *Telotaxis.*

True Prairie

The prairie grassland characterized by tall grasses (five to six or more feet tall) and mid-grasses (two to four feet tall) in the central part of the United States. cf. *Pampas.*

Truffle

Underground fungi in the genus *Tuber,* class Ascomycetes, or their fruiting bodies, edible.

Truncated Soil Profile

A *Soil profile* (q. v.) which has lost part or all of the *A* and *B horizons* (q. v.) by accelerated erosion or by cultivation.

Trypanosome

A flagellated *Protozoon* in the genus *Trypanosoma,* parasitic in the blood of various *Vertebrates,* and causing serious diseases in man and other animals such as sleeping sickness.

Tsetse Fly

A dipterous insect in the genus *Glossina,* sucks blood and transmits diseases caused by *Trypanosomes* such as sleeping sickness.

Tsunami

A high wave on shore areas, particularly bordering the Pacific Ocean, caused by an earthquake in the ocean floor.

Tuber

An enlarged, underground stem, tending to be oval or spherical in shape, usually rich in starch, and capable of vegetative reproduction of the plant, e.g., a potato tuber.

Tufa

A porous rock formed by the deposition of material, especially calcium carbonate from water, as in springs. cf. *Travertine, Sinter.*

Tularemia

A disease in rabbits, rodents, and man, caused by the microorganism *Pasteurella tularensis,* which is transmitted by insects.

Tullgren Funnel

A modification of the *Berlese funnel* (q. v.) for separating *Collembolons* (q. v.), *Mites,* larvae, and other small organisms from the soil. cf. *Baerman funnel.*

Tundra

The treeless land in arctic and alpine regions, varying from bare area to various types of vegetation consisting of grasses, sedges, forbs, dwarf shrubs, mosses, and lichens.

Turbidity

The condition of a body of water that contains suspended material such as clay or silt particles, dead organisms or their parts, or small living plants and animals.

Turf

The layer of low, dense grassland, comprising the above-ground portions and the upper roots and rhizomes with attached soil particles. cf. *Sward, Sod.*

Turgid

The condition of a cell or a tissue when it is swollen with water causing *Turgor pressure* (q. v.).

Turgor Pressure

The actual pressure of the sap within a cell against the cell wall resulting from the intake of water by *Osmosis* (q. v.).

Turion

A winter bud on some water plants that becomes detached, overwinters, and under favorable conditions develops into a new plant.

Turnover

(1) The mixing of layers of water in lakes in the spring and autumn. cf. *Thermal stratification.* (2) The period of time required for an organism to grow, mature, die, and undergo decomposition.

Tussock

A plant-form that is tufted, bearing many stems arising as a large, dense cluster from the crown, e.g., a large bunch grass such as Arizona fescue.

Twin Communities

Communities that are similar in a dominant or combining layers, or in *Synusiae* (q. v.), but vary in others.

Tychocoen

See *Ubiquist.*

Type

(1) A kind of vegetation, e.g., *Community-type, forest type,* birch type. (2) The one or more specimens of a species, subspecies, or variety on the basis of which the *Taxon* was described. (3) One of the groups of soils in a soils series, e.g., Miami silt loam type in the Miami series.

Typhoon

A *Tropical cyclone* or *Hurricane* (q. v.) in the Far East, particularly in the China Seas.

U

Ubiquist (Ubiquitist)
An organism that flourishes in several kinds of communities or ecosystems, e.g., the raven, red maple tree.

Ubiquitous
Refers to a *Ubiquist*.

Ultrasonic
See *Supersonic*.

Ultraviolet Radiation
The electromagnetic waves not perceptible to the human eye, between violet light waves and X-rays, from about 390 mu to 10 mu in length.

Unconformity
An irregular line of contact between two geological strata caused by exposure to erosion of the lower one before submersion and consequent deposition of the second stratum.

Underdispersion
See *Hypodispersion*.

Undergrazing
An intensity of grazing by livestock in which the forage

available for consumption under good management practices is not used sufficiently, thus causing loss of forage.

Undergrowth

Collectively the shrubs, sprouts, seedling and sapling trees, and all herbaceous plants in a forest.

Underpopulation

A size of population so low in number of individuals that mortality is greater than in denser populations largely because of increased exposure to unfavorable environmental conditions.

Undershrub

A low shrub.

Understocked

Refers to a range area on which a smaller number of livestock is present than it is capable of supporting adequately for a given season. cf. *Fully stocked, Overstocked.*

Understory

Collectively the trees in a forest below the upper canopy cover. cf. *Overstory.*

Uneven-aged

Refers to a forest in which considerable differences in the ages of trees occur.

Ungulate

A *Mammal* with hooves, e.g., horse, cow, swine, elephant.

Unicellular

Refers to an organism that consists of only one cell, e.g., blue-green algae, *Protozoans.*

Union

A homogeneous grouping of plant species within a given stratum or of the same or closely similar life-forms. cf. *Synusia.*

Uniparous

Refers to an animal that produces only one egg or one offspring at one time.

Unisexual

Refers to an organism that is either male or female. cf. *Hermaphrodite.*

Univoltine

Refers to an organism that has only one generation in a year. cf. *Multivoltine, Diapause.*

Unpalatable

Refers to plants and other kinds of food that are not readily eaten by animals.

Unspecialized

Not *Specialized* (q. v.).

Upper Austral Life Zone

One of the divisions in Merriam's *Austral life zone* (q. v.).

Upper Sonoran Life Zone

See *Sonoran life zone.*

Use, Actual

The total number and period of time livestock graze on a range, usually expressed as animal-unit months, cow months, or sheep months.

Use, Common

The practice of grazing a given range by more than one kind of livestock within the same grazing year.

Use, Dual

The grazing of a range area by more than one kind of livestock at the same time, such as cattle and sheep.

Use, Proper

The utilization of a *Range* so that the *Condition* is maintained in a good to excellent rating.

V

Vacuolated

Refers to a cell that contains one or more *Vacuoles.*

Vacuole

A space within a cell, enclosed by a membrane and containing a watery solution, the cell sap, surrounded by protoplasm.

Vacuole, Contractile

A structure in some one-celled organisms that excretes water by means of energy supplied by the *Cytoplasm.*

Vagility

The capability of an organism for *Dispersal* (q. v.).

Valence, Ecological

See *Ecological amplitude.*

Valley Train

Materials carried beyond a glacial ice-front by streams of melt water and deposited over a narrow area within a valley.

Van't Hoff Rule

See Q_{10}.

Vapor Pressure Deficit

The difference between the actual vapor pressure in the atmosphere in a certain space and the vapor pressure at saturation.

Variant

An organism, community, types of soil, etc., that differs sufficiently in its attributes from the typical specimen or norm to be classified as a variation of the group as a whole.

Variate

The attribute or characteristic such as height and weight that is used in statistical measurements of *Variation.*

Variation

Divergences in the characteristics of organisms, or other objects, of the same kind caused either by the environment or by differences in the genetic constitution of the organism. cf. *Phenotype, Genotype.*

Variegation

The irregular occurrence of patches, bands, or other areas on the surface of organs of plants and animals, such as leaves, caused chiefly by the lack of pigment in the cells of these areas or beneath them.

Variety

A taxonomic group or *Taxon* (q. v.) within a *Species,* or a *Subspecies,* e.g., *Juniperus communis* L. var. *sibirica* (Burgsd.) Rydb. cf. *Cultivar.*

Varve

A layer in a mass c lacustrine sediments, which may consist of coarser and finer sediments, deposited annually in a lake or sea.

Vascular

Refers to vessels or ducts that conduct fluids in organisms.

Vascular Bundle

A strand consisting of *Xylem* (q. v.) and *Phloem* (q. v.) in plants.

Vascular Cylinder

A vascular bundle with associated tissues in stems and roots of plants. cf. *Stele.*

Vascular Plant

A plant in the phylum Tracheophyta which includes the pteridophytes (ferns and their allies) and the spermatophytes (seed plants).

Vector

An organism, usually an insect, that transmits a pathogenic virus, bacterium, protozoon, or fungus from one organism to another, e.g., *Tsetse fly* (q. v.).

Vegetable Ball

A more or less spherical mass of plant material consisting of algae, other water plants, needles of trees, etc., formed by wave action in shallow water on sandy shores.

Vegetal

See *Vegetative.*

Vegetation

Plants in general, or the sum total of plant life in an area. cf. *Flora, Floristic, Community.*

Vegetational

Refers to *Vegetation,* in contrast to *Vegetative* (q. v.).

Vegetation (Vegetational) Cover

The sum total of plants and plant materials such as leaves, stems, and fruits that forms coverage on the surface of the soil; sometimes used in a more restricted sense to designate the sum of living plants on an area.

Vegetation (Vegetational) Type

A kind of *Vegetation* (q. v.) or the kind of *Community* (q. v.) of any size, rank, or stage of *Succession.*

Vegetative (Vegetal)

Refers to the nutritive and growth functions or structures of plants in contrast to the reproductive functions or structures; not to be confused with *Vegetation* or *Vegetational* (q. v.). cf. *Somatic.*

Vegetative Propagation

The propagation, or increasing the number, of plants by the use of *Vegetative* parts such as *rhizomes* (q. v.), *runners* (q. v.), *gemmae,* or other parts. cf. *Asexual reproduction.*

Vein

(1) See *Vascular bundle.* (2) A *Vessel* in animals that carries blood from the capillaries to the heart. (3) A thickened structure that gives support to wings of insects.

Velamen

A tissue in the outer part of aerial roots of certain plants, especially orchids, that absorbs water rapidly.

Veld (Veldt)

A tract of open country in South Africa occupied by grasslands at the higher elevations and by *Scrub* or *Savanna* at the lower elevations.

Venation

The arrangement of *Veins* in a leaf blade or in a wing of insects.

Verano

The long, dry period of the year in tropical America.

Vermiform

Refers to a shape of an object that is similar to a worm.

Vernal

Refers to spring. cf. *Aspection.*

Vernalization

The process of hastening the flowering phase of plants by subjecting young seedlings or other parts to low temperature, less commonly to a high temperature.

Versant

The general slope of a mountain range or a landscape.

Vertebrate

An animal in the subphylum Vertebrata, phylum Chordata, e.g., mammals, fishes, reptiles, birds.

Vesperal

Refers to evening time. cf. *Crepuscular.*

Vessel

A series of cells forming a tube-like structure in the *Xylem* (q. v.) of plants, conducts water and substances in solution.

Vestigial

Refers to a structure, function, or behavioural act of an organism that has so decreased in importance during the course of evolution that only a trace remains, e.g., the vermiform appendix in man. cf. *Primitive.*

Viability

The capability of a seed, spore, egg, or other organ of a plant or animal to continue or resume growth when it is exposed to favorable environmental conditions. cf. *Dormancy.*

Viable

Refers to the state of being alive. cf. *Viability.*

Vicariad (Vicarious Species)

One of a pair of closely related species, variety, or other *Taxon* that replace each other geographically.

Vicariation

The phenomenon of ecologically equivalent species, or taxonomically corresponding species, replacing (or "substituting") each other in similar environments in different geographic areas, e.g., caribou in North America and reindeer in Eurasia.

Vicinism

The condition of variation in a population or in an individual resulting from growing in close proximity to other organisms.

Virgin (Forest, Community, Region, etc.)

Refers to objects or aggregations, especially vegetation, essentially uninfluenced by human activity.

Virology

The branch of biology dealing with viruses.

Virus

A submicroscopic parasite in organisms consisting of nucleic acid and protein, incapable of increasing in number outside of the host cell, causing various diseases in plants and animals.

Vitalism

The doctrine that life processes are caused by some force that cannot be measured, in addition to the operation of the laws of chemistry and physics.

Vitality

The condition of vigor of organisms; the capacity to live and complete the life cycle. Braun-Blanquet classified plants according to states of vitality into four categories.

Vitamins

Organic substances required in minute quantities by plants and animals in their metabolic processes. cf. *Thiamine*.

Viticulture

The cultivation and production of grapes.

Viviparous

(1) Refers to an animal in which the embryo develops within its body and which produces living offspring, e.g., most mammals. cf. *Oviparous.* (2) Refers to a plant in which the embryo within the *Ovary* continues development without interruption, e.g., the mangrove; or the production of *Bulbils* or small plants instead of flowers and seeds, e.g., bulbous bluegrass.

Volume Weight

A figure denoting the number of times heavier a dry soil, including the pore space, is than an equal volume of water. cf. *Bulk density.*

W

Wadi (Wady)
A watercourse in deserts, dry except after rains, term used in southwest Asia and the Sahara. cf. *Arroyo, Wash.*

Wallace's Line
The line established by A. R. Wallace (1860) as a boundary between the *Oriental* and the *Australian Faunal regions* (q. v.).

Warm-bloodedness
See *Homoiotherm.*

Warm Front
The border between a mass of warm air advancing into or above a mass of colder air. cf. *Cold front.*

Warning Coloration
See *Aposematic.*

Wash
In southwestern United States, a dry bed of an intermittent stream, usually sandy and gravelly.

Wasting Disease
A disease of eelgrass (*Zostera marina*) often producing serious epidemics, caused by a *Slime mold* (*Labyrinthula* sp.).

Water Gap

A narrow valley or gorge in a ridge of mountains or hills, eroded by a stream, e.g., the Delaware Water Gap.

Water-holding Capacity

The amount of water, stated as the percentage of oven-dry soil, that is retained by the soil after the gravitational water has drained off. cf. *Field capacity.*

Waterlogged

The condition of a soil in which all the pore spaces are filled with water.

Water Requirement

The ratio of the number of units of water absorbed by a plant during the growing season to the number of units of dry matter produced by the plant during the same time. cf. *Transpiration coefficient.*

Watershed

(1) The total area of land above a given point on a waterway that contributes run-off water to the flow at that point. (2) A major subdivision of a drainage basin.

Waterspout

A tornado-like vortex and cloud occurring over a body of water.

Water Spreading

The application by means of stream diversion or otherwise of water over the land in order to increase the soil moisture supply for the growth of plants or to store it underground for subsequent withdrawal by pumping.

Water Table

The upper surface of the free ground water in a zone of saturation, except where it is separated by an underlying body of ground water by unsaturated material.

Water Table, Perched

The upper surface of a body of free ground water in a zone of saturation, separated by unsaturated material from another body of ground water in a saturated zone beneath. cf. *Perched water.*

Weather

The state of the atmosphere at any given time with regard to precipitation, temperature, humidity, cloudiness, wind movement, and barometric pressure. cf. *Climate.*

Weathering

The process of the physical and chemical disintegration of rocks and minerals.

Weed

A general term for any troublesome or otherwise undesirable plant, usually introduced, grows without intentional cultivation.

Wegener's Hypothesis

See *Continental drift.*

Weir

A dam across a water channel for diverting or for measuring the flow of water.

Westerlies

Winds that blow prevailingly from the southwest in the northern hemisphere, from the northwest in the southern hemisphere, located between the high pressure areas of the *Subtropics* and the arctic or antarctic circles.

White Alkali

See *Saline soil.*

Wilderness Area

See *Natural area.*

Wildlife

Collectively the non-domesticated vertebrate animals, except fishes, such as deer, moose, birds, etc.

Wildling

A seedling or a young plant that grew under natural conditions, not cultivated, outside of a nursery, and that has been dug for use as planting stock.

Williwaw

A sudden blast of wind descending from mountains to the sea, especially in the Straits of Magellan.

Willy-willy

A violent storm of rain and wind on the northwest coast of Australia; also applied in some parts of Australia to a local *Dust whirl* (q. v.).

Wilting

The temporary or transient loss of turgidity in a plant caused by a rate of transpiration in excess of the rate of absorption of water. Permanent wilting: wilting to such a degree that plants do not recover unless water is added to the soil soon after wilting occurs. Permanent wilting percentage (wilting-point, wilting-coefficient): the water remaining in the soil in percentage of dry weight of the soil when the plants are in a condition of permanent wilting.

Windbreak

A planting of trees and shrubs, usually in three or more rows to serve as a barrier to reduce or check the velocity of the wind. cf. *Shelterbelt.*

Winter Annual

A plant that germinates in the autumn, lives through the winter as a small plant usually, renews growth in the spring, flowers, produces fruit, and then dies.

Wireworm

The larva of certain slender beetles as in the genus *Agriotes.*

Witches' Broom

The abnormal brushlike production of numerous weak shoots toward the tip of a branch of a tree or a shrub, caused by a fungus or a mite.

Woodland

Any land used for the growth of trees and shrubs such as permanent woodland cover, plantings along roadsides and stream banks, *Shelterbelts,* farm *Woodlots,* etc.

Woodlot

A small area of land occupied by trees.

Working Depth (Roots)

The depth in the soil to which a large number of roots of a plant penetrate. cf. *Depth, effective soil.*

X

Xenia

An effect produced in the offspring in the endosperm of a seed brought about by the fusion of one of the sperms with the *Diploid* fusion nucleus in the *Ovule*.

Xenogamy

See *Cross-pollination*.

Xerarch

Refers to a successional sequence (*Sere*) which begins in a dry habitat. cf. *Xerosere, Hydrarch*.

Xeric

Refers to a dry habitat. cf. *Xerocolous, Xerophyte, Hygric, Mesic*.

Xerochase

A seed pod that opens in dry air and closes in moist air, e.g., carrot seed pods.

Xerocolous

Refers to animals living in dry places. cf. *Xerophilous, Hygrocole*.

Xeromorphy

The structure or form characteristic of organs of *Xerophytes* (q. v.), e.g., tough, leathery leaves on some desert shrubs.

Xerophilous, Xerophytic

Refers to a plant that is capable of growing in dry places, e.g., cactus. cf. *Xerophyte, Hydrophilous, Mesophyte.*

Xerophyte

A plant that can grow in dry places, e.g., creosote bush, cactus. cf. *Xerophilous.*

Xeroplastic

Refers to characteristics which are developed under the influence of drought.

Xerosere

A series of successional stages beginning in a dry area. cf. *Hydrosere, Sere, Xerarch.*

Xerothermic

Refers to a dry and warm climatic period, e.g., one of the postglacial periods.

X-rays

Electromagnetic rays, 0.1 to 50 mu, shorter than *Ultraviolet* and longer than *Gamma rays.*

Xylem

Woody tissue in the *Stele* (q. v.) of plants, conducts water and substances in solution.

Xylophagous

Refers to organisms that consume wood.

Y

Yeasts

One-celled plants in the class Ascomycetes, phylum Eumycophyta, the true fungi, reproduce vegetatively by budding, and convert sugar to alcohol and carbon dioxide in *Anaerobic respiration.*

Yield

The part of the production or *Productivity* (q. v.) of a group of organisms that is removed or expected to be removed by man, e.g., the number of deer killed during a hunting season, the timber produced by a stand of trees.

Yield Table

A table showing the volumes of timber that a stand of trees will produce at different ages (usually in ten-year periods) per unit of area.

Young Growth

See second growth.

Z

Zonal

Refers to *Zone*.

Zonal Soil

A kind of soil that has a permanent type of profile, characteristic of the prevailing conditions of the climate and vegetation, e.g., *Chernozem* (q. v.). cf. *Intrazonal soil.*

Zone

(1) Vegetation occurring in more or less well marked belts or areas much longer than wide, usually fairly uniform in physiognomy, as along lake shores, mountain sides, and sea shores. (2) One of the five great climatic belts of the earth; the two frigid zones, two temperate zones, and the torrid zone.

Zoochore

An organism that is normally disseminated by an animal. cf. *Diaspore.*

Zoogeography

The science that deals with the geographic distribution of animals. cf. *Biogeography, Plant geography.*

Zoology

The study of animals.

Zoophagous

Refers to organisms that feed on substances of animal origin.

Zooplankton

Animals occurring in *Plankton* (q. v.).

Zoospore

A motile spore, possessing one or more flagella, in certain algae and fungi.

Zygomorphy

See *Bilateral symmetry.*

Zygote

The product resulting from the union of two gametes; the fertilized egg.

REFERENCES

Alee, W. C., A. E. Emerson, O. Park, T. Park, and K. P. Schmidt. *Principles of Animal Ecology,* W. B. Saunders & Co., Philadelphia, Pa., 1949.

Andrewartha, H. G., and L. C. Birch. *The Distribution and Abundance of Animals,* University of Chicago Press, Chicago, Ill., 1954.

Becking, R. W. *The Zürich-Montpellier School of Phytosociology,* Botanical Review, vol. 23 (1957), pages 411-488.

Braun-Blanquet, J. *Plant Sociology,* McGraw-Hill Book Co., Inc., New York, N. Y., 1932.

Brown, Dorothy. *Methods of Surveying and Measuring Vegetation,* Bulletin 42, Commonwealth Bureaux Pastures and Field Crops, Hurley, Berks, England, 1954.

Cain, S. A. *Foundations of Plant Geography,* Harper and Bros., New York, 1944.

Cain, S. A., and G. M. de Oliveira Castro. *Manual of Vegetation Analysis,* Harper and Bros., New York, N. Y., 1959.

Clarke, G. L. *Elements of Ecology,* John Wiley & Sons, Inc., New York, N. Y., 1954.

Clements, F. E. *Plant Succession,* Carnegie Institute of Washington Publication 242 (1916), Washington, D. C.

Clements, F. E., and V. E. Shelford. *Bioecology,* John Wiley & Sons, New York, N. Y., 1939.

Dansereau, P. *Biogeography. An Ecological Perspective,* The Ronald Press, New York, N. Y., 1957.

Darlington, P. J., Jr. *Zoogeography: the Geographical Distribution of Animals,* John Wiley & Sons, New York, N. Y., 1957.

Daubenmire, R. F. *Plants and Environment, a Textbook of Plant Autecology,* John Wiley & Sons, Inc., New York, N. Y., 1959. 2nd Edition.

Dayton, W. A. *Glossary of Botanical Terms Commonly Used in Range Management,* U. S. Dept. of Agriculture Miscl. Publication No. 110 Revised (1950).

Dice, Lee R. *Natural Communities,* University of Michigan Press, Ann Arbor, Mich., 1952.

Dodson, E. D. *Evolution: Process and Product,* Reinhold Publishing Corporation, New York, N. Y., 1960.

Ecological Society of America. *Reports 1, 2, 3, of the Committee on Nomenclature,* (1933, 1934, 1935). (Mimeographed).

Eggleton, F. E. *Report of the Committee on Nomenclature of the Ecological Society of America,* Revised. (1952). (Photoprinted).

Esau, Katherine. *Anatomy of Seed Plants,* John Wiley & Sons, Inc., New York, N. Y., 1960.

Giese, A. C. *Cell Physiology,* W. B. Saunders Co., Philadelphia, Pa., 1957.

Good, R. *The Geography of Flowering Plants,* 2nd Edition, Longmans, Green & Co., London, England, 1953.

Hanson, Herbert C. *Ecology of the Grassland II,* Botanical Review, vol. 16 (1950), pp. 283-360.

Hanson, Herbert C., and E. D. Churchill. *The Plant Community,* Reinhold Publishing Corporation, New York, N. Y., 1961.

Hawley, A. H. *Human Ecology,* The Ronald Press Co., New York, N. Y., 1950.

Jackson, B. D. *A Glossary of Botanic Terms With Their Derivation and Accent,* Lippincott, Philadelphia, Pa., 1916.

Macfadyen, A. *Animal Ecology, Aims and Methods,* Sir Isaac Pitman & Sons, Ltd., London, England, 1957.

Mayr, E. (Editor). The Species Problem (A Symposium), American Association for the Advancement of Science Publ. No. 50, Washington, D. C., 1957.

McDougall, W. B. *Plant Ecology,* 4th Edition, Lea & Febiger, Philadelphia, Pa., 1949.

Meyer, B. S., D. B. Anderson, & R. H. Böhning. *Introduction to Plant Physiology,* D. Van Nostrand Co., Inc., Princeton, N. J., 1960.

Odum, E. P. *Fundamentals of Ecology,* 2nd Edition, W. B. Saunders Co., Philadelphia, Pa., 1959.

Oosting, H. J. *The Study of Plant Communities,* 2nd Edition, W. H. Freeman & Co., San Francisco, Calif., 1956.

Pearse, A. S. *Animal Ecology,* 2nd Edition, McGraw-Hill Book Co., Inc., New York, N. Y., 1939.

Sampson, A. W. *Range Management, Principles and Practice,* John Wiley and Sons, Inc., 1952.

Scott, J. P. *Animal Behaviour,* The University of Chicago Press, Chicago, Illinois, 1958.

Soil Conservation Society of America, *Soil and Water Conservation Glossary,* Journal of Soil and Water Conservation, vol. 7 (1951-1952), pp. 1-37.

Stoddart, L. A. *Range Management,* McGraw-Hill Book Co., Inc., New York, N. Y., 1943.

Swanson, C. P. *The Cell,* Prentice-Hall, Inc., Englewood Cliffs, N. J., 1960.

Tansley, A. G. *The British Islands and Their Vegetation,* 2 vols. Cambridge University Press, Cambridge, England, 1949.

U. S. Department of Agriculture. *Soils, the Yearbook of Agriculture, 1957,* U. S. Government Printing Office, Washington, D. C., 1957.

U.S. Soil Survey Staff. *Soil Survey Manual,* U.S. Government Printing Office, Washington, D.C., 1951.

Weaver, J. E., and F. E. Clements. *Plant Ecology,* 2nd Edition, McGraw-Hill Book Co., Inc., New York, N. Y., 1938.

Welch, Paul S. *Limnology,* McGraw-Hill Book Co., Inc., New York, N. Y., 1935.

Woodbury, A. M. *Principles of General Ecology,* McGraw-Hill Book Co., Inc. New York, N. Y., 1954.